CONSTRUCTION MATERIALS AND ENGINEERING

SHIPS AND SHIPBUILDING

TYPES, DESIGN CONSIDERATIONS AND ENVIRONMENTAL IMPACT

CONSTRUCTION MATERIALS AND ENGINEERING

Additional books in this series can be found on Nova's website
under the Series tab.

Additional e-books in this series can be found on Nova's website
under the e-book tab.

MECHANICAL ENGINEERING THEORY AND APPLICATIONS

Additional books in this series can be found on Nova's website
under the Series tab.

Additional e-books in this series can be found on Nova's website
under the e-book tab.

CONSTRUCTION MATERIALS AND ENGINEERING

SHIPS AND SHIPBUILDING

TYPES, DESIGN CONSIDERATIONS AND ENVIRONMENTAL IMPACT

JOSÉ A. OROSA
EDITOR

New York

Copyright © 2013 by Nova Science Publishers, Inc.

All rights reserved. No part of this book may be reproduced, stored in a retrieval system or transmitted in any form or by any means: electronic, electrostatic, magnetic, tape, mechanical photocopying, recording or otherwise without the written permission of the Publisher.

For permission to use material from this book please contact us:
Telephone 631-231-7269; Fax 631-231-8175
Web Site: http://www.novapublishers.com

NOTICE TO THE READER

The Publisher has taken reasonable care in the preparation of this book, but makes no expressed or implied warranty of any kind and assumes no responsibility for any errors or omissions. No liability is assumed for incidental or consequential damages in connection with or arising out of information contained in this book. The Publisher shall not be liable for any special, consequential, or exemplary damages resulting, in whole or in part, from the readers' use of, or reliance upon, this material. Any parts of this book based on government reports are so indicated and copyright is claimed for those parts to the extent applicable to compilations of such works.

Independent verification should be sought for any data, advice or recommendations contained in this book. In addition, no responsibility is assumed by the publisher for any injury and/or damage to persons or property arising from any methods, products, instructions, ideas or otherwise contained in this publication.

This publication is designed to provide accurate and authoritative information with regard to the subject matter covered herein. It is sold with the clear understanding that the Publisher is not engaged in rendering legal or any other professional services. If legal or any other expert assistance is required, the services of a competent person should be sought. FROM A DECLARATION OF PARTICIPANTS JOINTLY ADOPTED BY A COMMITTEE OF THE AMERICAN BAR ASSOCIATION AND A COMMITTEE OF PUBLISHERS.

Additional color graphics may be available in the e-book version of this book.

Library of Congress Cataloging-in-Publication Data

Ships and shipbuilding : types, design considerations and environmental impact / editor, Jose A. Orosa (Department of Energy, University of A Coruqa, Coruqa, Spain).
    pages cm
 Includes index.
 ISBN: 978-1-62618-787-0 (hardcover)
 1. Naval architecture. 2. Ships--Environmental aspects. I. Orosa Garcma, Jose A., editor of compilation.
 VM156.S54 2011
 623.8'1--dc23
                                2013015683

Published by Nova Science Publishers, Inc. † New York

CONTENTS

Preface   **vii**

Section 1: Ship Types and Design Considerations   **1**

Subsection 1: Design of Lines and Machinery Selection   **3**

Chapter 1   Parametric Design of Small Ships with the Use of NURBS Surfaces   **5**
Francisco Pérez-Arribas

Chapter 2   Description of a Propulsion System of a Ship Designed to Transport
LNG with a Steam Plant   **47**
*A. De Miguel Catoira, J. Romero Gómez, M. Romero Gómez
and R. Bouzón Otero*

Chapter 3   Regression Methods to a Gain Scheduling PID Controller to Guarantee
the Automatic Steering of Ships   **63**
*Héctor Quintián Pardo, José Luis Calvo-Rolle,
José Luis Casteleiro-Roca and José Antonio Orosa Garcia*

Chapter 4   Knowledge Model Approach Based in Rules for TACAN Air Navigation
System   **85**
*Xosé Manuel Vilar Martínez, Juan Aurelio Montero Sousa,
José Luis Calvo-Rolle, José Luis Casteleiro-Roca
and José Antonio Orosa García*

Subsection 2: Safety and Health Risk   **103**

Chapter 5   Research about the Risk of Explosion on Board from Using Liquefied
Natural Gas as Fuel   **105**
Saturnino Galán, José A. Orosa, Angel Rodríguez and José A. Pérez

Chapter 6   Research About Safety of Wire Rope on Board   **125**
Saturnino Galán, José A. Orosa, Angel Rodríguez and José A. Pérez

Section 2: Ships and Environmental Impact   **151**

Subsection 3: The International Maritime Organization (IMO)   **153**

Chapter 7   Research about the New IMO Convention   **155**
*Rebeca Bouzón, Ángel M. Costa, A. De Miguel Catoira,
J. Romero Gómez and M. Romero Gómez*

Chapter 8	IMO Standard and Gas Emissions Reduction from Ships *Ángel M. Costa, Rebeca Bouzón, A. De Miguel Catoira,* *J. Romero Gómez and M. Romero Gómez*	**167**

Subsection 4: Port Ship Emissions   **185**

Chapter 9	Air Quality Impact Assessment of In-Port Ship Emissions: Methodological Issues and Case-Study Examples *Giovanni Lonati*	**187**
Chapter 10	Applying the Life Cycle Thinking to Sea Ports: The Case of a Slovenian Commercial Port *Boris Marzi, Stefano Zuin, Gregor Radonjič and Klavdij Logožar*	**205**
Index		**219**

PREFACE

Some authors define a ship as a complex vehicle that must be self-sustaining and, in consequence, must be constructed in accordance with different design criteria. On the basis of this point of view, this book was organised in two main sections titled "Section 1: Ship types and design considerations", and "Section: 2. Ships and environmental impact".

Owing to the multitude of ship types and design considerations, this can be analysed based on different engineering aspects like the design of ship lines, crew members' accommodation, and health facilities, and more subsections were incorporated within the first section.

In accordance with this initial classification, the first subsection, titled "Design of lines and machinery selection", shows four chapters as clear examples of design criteria based on engineering aspects. The first chapter, Chapter 1, is an example of design lines of small ships. Chapter 2, is an example of machinery selection based on energetic criteria and, finally, Chapters 3 and 4 show new models for automatic steering of ships. These four chapters are a clear example of how complex ship designing can be, and which are the new research topics in these areas.

On the other hand, as was commented earlier, the second subsection, "Safety and health risk", shows other kinds of ship design criteria centred on crew members and health. Towards these, new investigations gave examples about risky situations on board that have not been analysed in depth, like for example, risk of explosions (Chapter 5), and wire rope conditions (Chapter 6).

In the second section, "Ships and environmental impact", two more subsections can be seen. The first subsection, titled "The International Maritime Organization (IMO)" shows in Chapters 7 and 8 the new modifications of IMO and, in particular, the main characteristics of dismantling and scrapping of ships.

Finally, the last subsection titled "Port ship emissions", shows in Chapters 9 and 10, the effect of gas emissions from ships based on a theoretical approach, and gives real case studies.

A Coruña
Editor
January 13, 2013

SECTION 1: SHIP TYPES AND DESIGN CONSIDERATIONS

SUBSECTION 1: DESIGN OF LINES AND MACHINERY SELECTION

In: Ships and Shipbuilding
Editor: José A. Orosa

ISBN: 978-1-62618-787-0
© 2013 Nova Science Publishers, Inc.

Chapter 1

PARAMETRIC DESIGN OF SMALL SHIPS WITH THE USE OF NURBS SURFACES

Francisco Pérez-Arribas[*]
Naval Architecture School of Madrid (UPM), Madrid, Spain

ABSTRACT

This chapter presents a mathematical method to define a ship hull based on numerical constraints directly related to geometric features of the hull surface and on naval architecture parameters that uniquely define a hull form. These geometric parameters have physical, hydrodynamic or stability implications from the design point of view. B-spline curves and surface representations were combined with a constrained approach to produce the final hull display. The methods presented in this chapter can be applied in the design of round bilge or hard-chine hulls that represent most of small boat designs (motor yachts, sailing ships, small fishing vessels, recreational crafts and patrol boats).

The presented methods follows the traditional design principles of naval architecture, starting with a Sectional Area Curve that controls buoyancy and a waterplane curve that ensures stability in the case of round bilge hulls, or in the case of hard chine hulls, the method considers control curves such as the center, chine and sheer lines, as well as their geometric features including position, slope and, in the case of the chine, enclosed area and centroid. A B-spline outline of these curves is adopted and a nonlinear problem is solved to obtain their constrained definition that agrees with the geometric design parameters. Different boundaries are also introduced into the constrained definition that control the limits of the hull, the centre and deck lines, which are complemented with tangent values at these edges to gain local control of the hull. A net of curves or hull stations is created that matches the previously defined constraints. This is an important advantage of the method because a hull form library or template is not necessary, but limit the type of ship hulls that can be attained. The method continues with a constrained B-spline fitting of points on the stations that ensures the tangent angles and checks the distance from the points to the B-spline. A final lofting surface of the previous B-spline curves produces the hull surface. Different application examples for round bilge and hard chine hulls are presented at the end of the chapter.

---

[*] Email: Francisco.perez.arribas@gmail.com.

INTRODUCTION

The definition of a ship hull is one of the most restraining processes in initial ship design. A hull form is usually designed by modifying an existing hull (parent hull), which can include a direct manipulation of the surface control vertices. Although this has become a standard practice and produces good results, it necessitates a great deal of manual work because the designer has to manipulate individual control vertices, it does not allow for rapid creation or modification of the hull surface in the initial phases of the ship design process, and it cannot ensure important magnitudes of the ship hull such as displacement (the volume of the underwater portion of the ship) until a later calculation is made, and a trial and error procedure is sometimes necessary. Thus, a hull form should be generated as early as possible to provide demanded design parameters that are important for subsequent design stages such as stability, hydrostatics, propulsion, and cost analysis.

The constrained generation of a ship hull can ensure that design parameters are met. The design algorithm generates an appropriate hull constrained by the design parameters without further human interaction. However, these fixed formulations do not allow a great range of flexibility in the number of hull shapes to be generated.

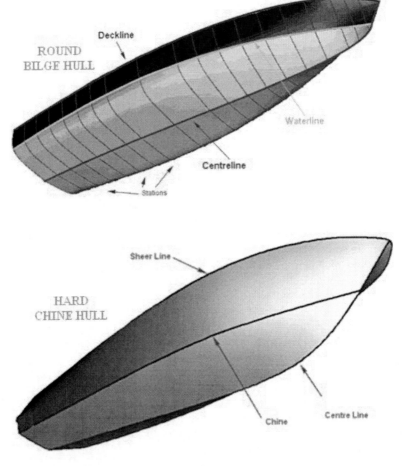

Figure 1. Example of ship hulls that can be designed.

In some cases, the constrained or parametric definition of ship hulls is beneficial during the initial phases of the ship design because it speeds up the creation process, provides more clarity about the influence of various parameters, and allows for the development of a range of concept hull surfaces. Today, computational optimisation techniques based on computational fluid dynamics (CFD) can be used together with systematic constraint definitions of a ship hull to produce faster, more comfortable and safer ships [1, 2]. These optimisation methods require complete geometric details of the design and correct management of hull information so that one can understand the relationship with the constrained design of a ship hull.

This chapter presents a constrained design method for defining a ship hull. As a consequence of the formulation of a constrained methodology, the design is limited to certain types of hulls. Nevertheless, the range of surfaces that can be generated make the method suitable for motor yachts, sailing ships, small fishing vessels and round bilge patrol boats. Practical examples will be used to show how the proposed method affords reliable hull design solutions in a very short time.

Figure 1 presents an example of a hull that can be automatically defined with the presented method. The method works different for round bilge (Figure 1 up) and hard-chine hulls (Figure 1 down). The first kind of ships is designed using a Sectional Area Curve (SAC) and the Waterplane definition, whilst the second group is designed based on the geometry of the main curves of the hull: chine, sheer and center lines.

1. BACKGROUND

Parametric design methodologies can be traced back to Kuiper [3], who generated hull shapes in the 1970s using conformal mapping techniques from hull parameters, rather than from offset hull data points. Kuiper generated hulls by constructing different waterline polynomials according to coefficients controlled by draught functions. The hull representation techniques of that time did not allow the development of hull shapes in a convenient and accurate way.

Reed and Nowacki [4] developed a compromise between polynomials and conformal mapping techniques; namely, polynomials were used to represent the hull above the waterline, and conformal mapping techniques were used for the underwater part of the ship. B-splines and NURBS functions were used by Creutz and Shubert [5] for ship hull design. They developed a procedure to generate B-spline curves from form parameters. These early studies demonstrated how NURBS and B-splines could adequately represent the geometry of a ship hull.

Keane [6] developed simple hulls using constrained generation techniques based on conformal mapping and studied the influence of certain parameters on ship stability. This is one of the first optimization procedures to be based on parameter variation.

Yilmatz and Kukner [7] determined hull stations parametrically by applying a regression technique to a large database of hulls. This method can only be used for certain fishing vessels because those ships were used to construct the database. Mancuso [8] used parametric methods to define sailing ship hulls.

Different authors, such as Kim [9] and Bole [10], researched this topic by subdividing the ship hull into multiple domains, such as the entrance, flat bottom, and flat side. The proposed method of this chapter will subdivide hard-chine hulls into one domain below the chine line and one above the chine line. In the case that a spray rail is present, a third domain will be implemented.

Commercial hull design software packages now include modules for the basic constrained generation of ship hulls, demonstrating that designers have a clear need for such a technique.

These commercial software modules were reviewed by Bole in reference [10]. Some of the software packages use non-intuitive parameters such us curvatures, derivatives or numerical weights, and other packages are limited on the other end producing very simple hulls based just on main dimensions. In general, users tend to revert to manual manipulation of a parent hull from a software database, which has inherent limitations that have been discussed previously. Occasionally, the designer has to contact the software manufacturer to acquire a parent hull archive. A general overview of the advances in computational methods for ship hull definition during the last years is provided in [11].

The transformation of a parent hull is limited because of the numerical techniques that are used. These methods are based on relocating and scaling hull sections while maintaining the desired dimensions in a trial and error manner until a selected displacement and LCB are reached. Waterline modifications are generally not considered in these kinds of transformations.

The literature review suggests that a constrained definition of a ship hull can be successfully made by directly generating the hull surface from numerical parameters, as described in the present chapter, or by altering the shape of a given hull surface to match specified parameters. Using the second group of methods, the hull may be created using a direct or iterative procedure. The second group of methods is more flexible regarding the types of hulls that can be defined, but the templates cannot be defined or altered. Furthermore, these templates can be difficult to manage for a non-advanced user.

Constrained generation tools build the hull surface representation from relatively few parameters, and this information is used to determine the dimensions, shapes and volumetric properties of the hull. For this reason, this chapter uses an analytic definition of these curves for the hull stations instead of templates. These analytic curves restrain the range of ship hulls to those mentioned in the introduction. It is exceedingly difficult to develop mathematical procedures to describe a variety of different hull forms, and it is thus necessary to consider a different formulation for each hull type.

The use of the SAC is a common procedure found in the literature, but the definition of a constrained waterline is less common. As a difference with the present chapter, the constrained definition of the SAC and the waterline is based on an iterative minimisation of a weighted error function that depends on target values of area and the centre of gravity of the curves. The present chapter allows for the definition of such curves based on the solution to a nonlinear system of equations, as will be detailed in the following sections. The introduction of the tangent angles at the ends of the curves enables the designer to gain intuitive control over the area distribution. This control could be also gained including higher order moments of the curves into the problem definition, but these are less intuitive and would increase the complexity of the mathematical problem.

Based on the previous references, it can be observed that the use of parametric techniques in ship design is not unusual in the literature. However, the applicability of these techniques to hard-chine planing hulls is quite limited. One notable reference has been found by Calkins et al. [12], that defined a hard-chine hull using a parametric methodology, but it did not include B-spline modeling, which today is a standard in naval architecture; rather, it covered the hull shape with straight stations, which is unrealistic most of the time.

The use of parameters to study the hydrodynamic properties of hard-chine hulls is quite common, like in the significant papers in this field written by Savitsky [13-16]. Commercial hull design software packages now include modules for the basic, constrained generation of ship hulls, demonstrating that designers have a clear need for such a technique.

2. DESCRIPTION OF THE METHOD FOR ROUND BILGE HULLS

The primary design characteristics of any ship hull surface are that it should float and stay upright while accommodating cargo. Many shapes satisfy these requirements, including a simple box, but when movement is required, the shape requirements change as a result of the hydrodynamic effects of the water flowing around the hull. The SAC has a substantial impact on ship hydrodynamics and the internal volume distribution while the waterline (WL) controls the hull stability, which is also influenced by the internal weight distribution that cannot be considered geometrically. As such, these two curves have been selected as constraints for the presented method and their definition impact the entire ship design process.

The selected parameters used to define these curves have meaning for the ship designer and can be related to further aspects of ship design. There are several good textbooks and references that show the effect of most of the parameters that the presented method uses into the performance of the ship hull, such as the ones of Lamb [17] and Lewis [18]. The objective of the presented method is to create a B-spline surface to represent a round bilge hull based on the constraints displayed in Table 1. In the case of a round bilge hull, the method will first create a SAC to constrain displacement and its longitudinal centre of buoyancy that will match the ship's longitudinal centre of gravity when floating at rest. Imposing these constraints achieves the design equilibrium position of the ship.

A WL of a given area and centre of area is then defined. This is very important to ensure the initial stability of the ship, which can be calculated once the internal mass distribution of the ship is obtained. Both curves share the length at the waterline (Figure 1) and will be modelled as B-splines.

The ship's centreline is included as part of this method. This line is not defined with parameters like the SAC or waterline because there are not common naval design parameters associated to that line. A smooth 2D curve that sets the lowest points of the hull is imposed by the designer. The distribution of the tangent angles at points on the centreline (dead-rise angle or rise of floor) is also included, providing local control of the centre part of the ship. Different hull forms can be obtained by altering this distribution of angles: for example, fast boats have a dead-rise angle diminishing towards the stern (i.e., the initial part of the stations are straight segments) to gain hydrodynamic lift, while slow ships have round stations of zero dead-rise angle.

Table 1. Parameters for a round bilge hull

Constraint	Name	Meaning	Where?
Displacement	Disp	Submerged volume of the ship = weight / water density. Area enclosed by SAC	SAC
Longitudinal centre of buoyancy	LCB	Longitudinal position of the centre of gravity of the submerged volume	SAC
Abscissa of the maximum of SAC	XMAX	Longitudinal position of the maximum of SAC. Related to internal arrangement	SAC
Angles at the ends of the SAC	α_I, α_F	Controls volume distribution inside the ship	SAC
Waterlength	LWL	Length of the flotation	SAC/WL
Area at waterline	AWP	Area enclosed by the Waterline	WL
Longitudinal centre of the flotation area	LCF	Longitudinal position of the centre of gravity of the waterline	WL
Angles at the ends of WL	α'_I, α'_F	Controls area distribution in the WL	WL
Ship's draft	T	Height of the WL	Input
Centreline		Initial points of the stations	Boundary
Dead-rise angles	β_i	Tangent at initial points of sections. Local control.	Boundary
Deckline		Final points of the stations	Boundary
Angles at the deckline	δ_i	Tangent at the end of sections. Local control. Convexity.	Boundary

Correspondingly, the deckline and tangent angle distribution at this edge have been included as constraints. Note that the deckline can be a 3D curve. These curves are normally obtained from initial sketches of the ship's internal arrangement, as in Figure 1. These sketches can represent the form topology with a limited amount of information. As the sketch demonstrates, these curves are very effective at representing the shapes within the ship hull that the designer wants to control. Constraints on the curve definition can be used to accurately reproduce stations and can be manipulated without the need for the user to move the definition vertex. The stations on the hull create the framework where the hull surface will lean. Every station will enclose an area below the WL calculated from the SAC and will begin with a determined dead-rise angle from the corresponding constraint. At the waterline height (T), the station must have a given abscissa imposed by the waterline. The final point of the curve will be at the deckline, arriving there at a constrained angle. A direct B-spline model of these stations would not produce good results because obtaining the position of the control vertex by solving a non-linear system of equations would require extra boundary conditions. For example, if the start of the B-spline were forced to be a straight line then extra control points and corresponding conditions would be required to determine the system of equations. For this reason, an analytic solution is adopted for the underwater part of the stations. A curve can be computed that starts at the centreline with a given angle, which encloses a given area (from the SAC) and has a determined ordinate (from the WL) at the ship's draft. The part of the station between the waterplane and the deckline is computed with a 2nd order Bézier curve that matches the angles at its ends and that not require any other property except to be a soft line; so this is the most simple solution. Analytic curves cannot generally be used in CAD programs, so a constrained B-spline fitting of points of each station is performed. This fitting includes points from the underwater (analytic curves) and above-water (Bézier curves) part of every station, and will impose the endpoints of the curves and their accompanying tangent angles as described in Table 1. The adjustment can impose a degree and a number of control points, although in practice it is preferable to use cubic B-splines and increase the number of control points until a given construction tolerance is achieved. A final lofting or skinning of the previously defined B-splines will provide the final constrained surface of the ship's hull.

2.1. Constrained Definition of the SAC: Introduction to B-Splines Curves

The method begins by building the SAC of the ship hull as shown in Figure 2. This curve will enclose a given area (Displacement) and a centre of buoyancy (LCB) and is extended along the ship's waterlength (LWL). The longitudinal position of the maximum of the curve (XMAX) is also constrained. Positions and angles at the ends of the curve $α_I$ and $α_F$ produce the final set of equations. The inclusion of the angles in the definition of the SAC enables the designer to gain control over the volume distribution that the SAC effects: a higher angle increases the volume at this part of the ship. The distribution of area towards the ends of the curves is related to the prismatic coefficient, a well known non-dimensional term used in naval architecture. Although it is not directly included in the method, it can be affected by the variation of the angles at the ship ends.

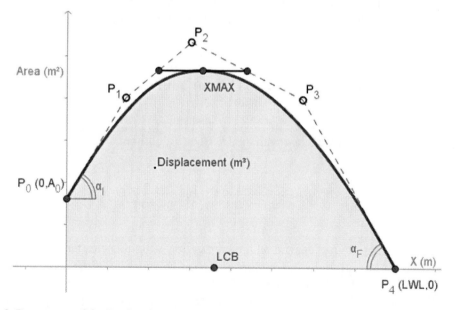

Figure 2. Parameters of the Sectional Area Curve (SAC).

This method computes the SAC as the solution of a nonlinear set of equations that form the 5 control points of a cubic B-spline, Eq. (1). These equations hold for all the design constraints shown in Figure 2.

The aspect of the SAC of any conventional ship without a central parallel body is similar to the one represented in this figure: a thin entrance to improve hydrodynamic performance and a soft run that enables undisturbed water flow. The area at the ends can be a non-zero value because the ship can have a bulbous bow and a transom stern. Bulbous bows cannot be modelled with the shape of sections used here.

$$\mathbf{SAC}(t) = \sum_{j=0}^{4} B_j^3(t) \cdot \mathbf{P}_j \qquad (1)$$

Where B_j^3 corresponds to the jth basis function of a 3rd degree B-spline that is calculated using de Boor's algorithm considering a uniform knot vector. To introduce the notation for this chapter, a brief review of B-splines follows. A B-spline curve is formed by several pieces of polynomial curves, called Bézier pieces, and the whole curve is C^2 (common curvature or second derivatives) at the junctions in the case of cubic B-splines.

The curve is defined with a polygon, called the control polygon, and with an interpolation algorithm that allows its construction to relate the curve to the control polygon. The interpolation steps are encoded in a family of piecewise polynomial functions, $B^n_j(u)$, called B-spline functions of nth degree, and are calculated using De Boor's algorithm [19]. Cubic B-splines are the most widely used curves in ship design and the ones that generally fit better the traditional loftsman's splines.

A B-spline curve, $s(u)$, is a linear combination of basis functions and $m + 1$ control points, P_j, as coefficients. Therefore, B-spline curves are parametric, $x = X(u)$, $y = Y(u)$, $z = Z(u)$, where the parameter u is considered inside $[0,1]$ in this chapter. In the 2D plane, $P_j = (Xj, Yj)$, $j = 0, \ldots, m$, generate a B-spline $s(u)$ of the nth degree,

$$s(u) = \sum_{j=0}^{m} P_j \cdot B_j^n(u) = (X(u), Y(u)) = \sum_{j=0}^{m} \left(X_j \cdot B_j^n(u), Y_j \cdot B_j^n(u) \right)$$

where the basis functions are obtained with the De Boor's algorithm:

$$B_j^0(u) = \begin{cases} 1 & u \in [u_{j-1}, u_j) \\ 0 & u \notin [u_{j-1}, u_j) \end{cases} \qquad B_j^n(u) = \frac{u - u_{j-1}}{u_{j+n-1} - u_{j-1}} \cdot B_j^{n-1}(u) + \frac{u_{j+n} - u}{u_{j+n} - u_j} \cdot B_{j+1}^{n-1}(u)$$

The basis functions, $B^n_j(u)$, depend on the knot vector, u_j. In this chapter, the knot vector has been selected to be uniform with a multiplicity equal to the order of the curve at its ends, where the order is defined as the degree $+ 1$.

In this manner, the B-spline interpolates the ends of its control polygon at $u = 0$ and $u = 1$, and it is tangent at its ends to the first and last segment of its control polygon. This last property simplifies the mathematical definition of the curves used in the method. Notice that the derivative, $s'(u)$, of a B-spline is a linear combination of the derivatives of the basis functions:

$$s'(u) = \sum_{j=0}^{m} P_j \cdot B_j'^n(u) = (X(u), Y(u)) = \sum_{j=0}^{m} \left(X_j \cdot B_j'^n(u), Y_j \cdot B_j'^n(u) \right)$$

When the number of control points ($m+1$) is equal to the order of the curve ($n+1$), the B-spline curve is formed with only one piece and can be called a Bézier curve of nth degree. In this chapter, all the curves will be referred to as B-splines for the sake of clarity. A variable presented in bold letters indicates that this element is formed from several components, such as points, vectors or B-spline curves that have an X, Y and Z component.

Figure 2 shows that the unknowns are the positions of the control points of this figure, $P_1(XP_1, YP_1)$, $P_2(XP_2, YP_2)$ and $P_3(XP_3, YP_3)$, because the ends are fixed by the constraints LWL and A_0. The maximum XMAX is placed at the joint of the two cubic Bézier curves that compound the cubic B-spline. This has a graphical meaning that is depicted in the figure according to the Casteljau algorithm. Considering a uniform knot vector for SAC(t), [0,0,0,0,0.5,1,1,1,1], XMAX will be placed at $t=0.5$. This means that the knot will be adjacent to the midpoint of the curve.

The conventional SACs of ships have maximum values near the middle of the ship's length, so the assumption for the parametric position of XMAX is a realistic approach that simplifies the mathematical definition of the problem and produces Eqs. (2) and (3)

$$\frac{XP_1 + 2 \cdot XP_2 + XP_3}{4} = XMAX \tag{2}$$

$$YP_1 = YP_3 \tag{3}$$

Eqs. (4) and (5) are easily obtained from the properties of the control polygon of a B-spline at its ends, which are tangent to the curve:

$$YP_1 = YP_0 + tg(\alpha_I) \cdot (XP_1 - XP_0) \tag{4}$$

$$YP_3 = YP_4 + tg(\alpha_F) \cdot (XP_4 - XP_3) \tag{5}$$

The most difficult part is the inclusion of the enclosed area and its centre of gravity into the definition of the problem. Although the area of the closed B-spline can be easily computed with Greens' theorems, in this case the area is enclosed by the curve and the X-axis. This can be solved by integration in the parametric domain:

$$Disp = Area = \int dA = \int_{XP_0}^{XP_4} Y(t) \, dX = \int_0^1 Y(t) X'(t) \, dt \tag{6}$$

For this particular case of a cubic B-spline with 5 control points, the expression of $X'(t)$ and $Y(t)$ are cubic polynomials computed with the cubic basis functions of the B-spline and their derivatives. The result of the integral of Eq. (6) can be expressed in matrix form as in Eq. (7):

$$[X] = \begin{pmatrix} XP_0 \\ XP_1 \\ XP_2 \\ XP_3 \\ XP_4 \end{pmatrix} \quad [Y] = \begin{pmatrix} YP_0 \\ YP_1 \\ YP_2 \\ YP_3 \\ YP_4 \end{pmatrix} \quad [\Phi] = \begin{pmatrix} \Phi_{00} & \cdots & \Phi_{04} \\ \vdots & \ddots & \vdots \\ \Phi_{40} & \cdots & \Phi_{44} \end{pmatrix} \quad \Phi_{ij} = \int_0^1 B_i^3(t) B_j'^3(t) dt$$

$$[\Phi] = \begin{pmatrix} -\dfrac{1}{3} & \dfrac{247}{2688} & \dfrac{19}{1680} & \dfrac{1}{2688} & 0 \\ \dfrac{-97}{672} & 0 & \dfrac{59}{1120} & \dfrac{31}{1120} & \dfrac{1}{672} \\ \dfrac{-1}{48} & \dfrac{-143}{2240} & 0 & \dfrac{143}{2240} & \dfrac{1}{48} \\ \dfrac{-1}{672} & \dfrac{-31}{1120} & \dfrac{-59}{1120} & 0 & \dfrac{97}{672} \\ 0 & \dfrac{-1}{2688} & \dfrac{-19}{1680} & \dfrac{-247}{2688} & \dfrac{1}{3} \end{pmatrix} \quad Disp = [X]^t \cdot [\Phi] \cdot [Y] \tag{7}$$

Expression (7) shows that the area will be a function of the cross product of the coordinates (XPi, YPi) of the control points; it is not a linear constraint as in the previous conditions (2)–(5).

The final constraint is the inclusion of the centre of buoyancy (LCB) into the problem definition. This can be computed as a solution of the following integral:

$$LCB = X_{cg} = \frac{\int X \, dA}{Disp} = \frac{\int_{XP_0}^{XP_4} X(t) Y(t) \, dX}{Disp} = \frac{\int_0^1 X(t) Y(t) X'(t) \, dt}{Disp} \tag{8}$$

As in the previous calculations of the enclosed area, expression (8) can be computed with the cubic basis functions and their derivatives, and can be expressed in matrix form as follows:

$$[X^2] = \begin{pmatrix} XP_0^2 \\ XP_1^2 \\ XP_2^2 \\ XP_3^2 \\ XP_4^2 \end{pmatrix} \quad [\dot{X}] = \begin{pmatrix} XP_4 \cdot XP_3 \\ \vdots \\ XP_4 \cdot XP_0 \\ XP_3 \cdot XP_2 \\ \vdots \\ XP_3 \cdot XP_0 \\ XP_2 \cdot XP_1 \\ XP_2 \cdot XP_0 \\ XP_1 \cdot XP_0 \end{pmatrix} \quad [\Omega] = \begin{pmatrix} \Omega_{00} & \cdots & \Omega_{04} \\ \vdots & \ddots & \vdots \\ \Omega_{40} & \cdots & \Omega_{44} \end{pmatrix} \quad \Omega_{ij} = \int_0^1 B_i^3(u) B_i^3(u) B_j'^3 du$$

$$[\Psi] = \begin{pmatrix} \Psi_{43}^0 & \cdots & \Psi_{40}^0 & \Psi_{32}^0 & \cdots & \Psi_{30}^0 & \Psi_{21}^0 & \Psi_{20}^0 & \Psi_{10}^0 \\ \vdots & \ddots & \vdots & \vdots & \ddots & \vdots & \vdots & \vdots & \vdots \\ \Psi_{43}^4 & \cdots & \Psi_{40}^4 & \Psi_{32}^4 & \cdots & \Psi_{30}^4 & \Psi_{21}^4 & \Psi_{20}^4 & \Psi_{10}^4 \end{pmatrix}$$

$$\Psi_{ij}^p = \int_0^1 B_p^3 \left[B_i^3(u) B_j'^3(u) + B_j^3(u) B_i'^3(u) \right] du$$

$$[\Omega] = \begin{pmatrix} -\dfrac{1}{2} & \dfrac{3}{80} & \dfrac{1}{10} & \dfrac{1}{80} & 0 \\[2mm] \dfrac{-3}{80} & 0 & \dfrac{9}{40} & \dfrac{3}{20} & \dfrac{1}{80} \\[2mm] \dfrac{-1}{10} & \dfrac{-9}{40} & 0 & \dfrac{9}{40} & \dfrac{1}{10} \\[2mm] \dfrac{-1}{80} & \dfrac{-3}{20} & \dfrac{-9}{40} & 0 & \dfrac{31}{80} \\[2mm] 0 & \dfrac{-1}{80} & \dfrac{-1}{10} & \dfrac{-31}{80} & \dfrac{1}{3} \end{pmatrix}$$

$$[\Psi] = \begin{pmatrix} 0 & 0 & 0 & 0 & \dfrac{5}{1344} & \dfrac{11}{1680} & \dfrac{1}{672} & \dfrac{71}{1344} & \dfrac{1}{48} & \dfrac{97}{672} \\[2mm] \dfrac{89}{13440} & \dfrac{17}{6720} & \dfrac{1}{2688} & 0 & \dfrac{27}{448} & \dfrac{31}{1120} & \dfrac{1}{13440} & \dfrac{143}{2240} & \dfrac{-47}{6720} & \dfrac{-247}{2688} \\[2mm] \dfrac{11}{240} & \dfrac{19}{1680} & \dfrac{1}{840} & 0 & \dfrac{59}{1120} & 0 & \dfrac{-1}{840} & \dfrac{-59}{1120} & \dfrac{-19}{1680} & \dfrac{-11}{240} \\[2mm] \dfrac{247}{2688} & \dfrac{47}{6720} & \dfrac{-1}{13440} & 0 & \dfrac{-143}{2240} & \dfrac{-31}{1120} & \dfrac{-1}{2688} & \dfrac{-27}{448} & \dfrac{-17}{6720} & \dfrac{-89}{13440} \\[2mm] \dfrac{-97}{672} & \dfrac{-1}{48} & \dfrac{-1}{672} & 0 & \dfrac{-71}{1344} & \dfrac{-11}{1680} & 0 & \dfrac{-5}{1344} & 0 & 0 \end{pmatrix}$$

$$Disp \cdot LCB = [Y]^t [\Omega] [X^2] + [Y]^t [\Psi] [\hat{X}] \tag{9}$$

Equation (9) is a function of the cross product of the control point coordinates and their squared and cubic powers. This yields six equations (2)–(5), (7) and (9) with six unknowns XP_1, XP_2, XP_3, YP_1, YP_2, and YP_3 which form a nonlinear system of equations that has to be solved. A direct approach to the solution of such a system can be numerically difficult to obtain because the solution is very sensitive to initial estimates of the solutions, and non-realistic results can be obtained because of the nonlinearity of Eqs. (8) and (9).

Manipulation of the linear conditions Eq. (2)–(5) enables further simplifications:

$$XP_3 = \frac{YP_4 - YP_0 + tg(\alpha_I) \cdot XP_0 + tg(\alpha_F) \cdot XP_4}{tg(\alpha_F)} - \frac{tg(\alpha_I)}{tg(\alpha_F)} \cdot XP_1 \quad \rightarrow f(XP_1)$$

$$XP_2 = 2 \cdot XMAX - \frac{1}{2} \cdot XP_3 - \frac{1}{2} \cdot XP_1 \quad \rightarrow f(XP_1)$$

$$YP_3 = YP_4 + tg(\alpha_F) \cdot (XP_4 - XP_3) \quad \rightarrow f(XP_3) \rightarrow f(XP_1)$$

Considering these simplifications and Eq. (4), the initial six-equation system is equivalent to a much simpler system (10) of two equations with two unknowns, XP_1 and YP_2:

$$\begin{aligned} & a \cdot XP_1^2 + b \cdot XP_1 + c \cdot YP_2 + d \cdot XP_1 \cdot YP_2 + e = Disp \\ & f \cdot XP_1^3 + g \cdot XP_1^2 \cdot YP_2 + h \cdot XP_1 \cdot YP_2 + j \cdot XP_1^2 + k \cdot YP_2 + l \cdot XP_1 + m \cdot XP_1 \cdot XP_2^2 + n = \\ & = Disp \cdot LCB \end{aligned} \tag{10}$$

Where a, b, c,...n are constants that depend on the initial numerical constraints. The solution of this nonlinear system is calculated with a Powell hybrid algorithm. This algorithm is a variation of Newton's method, which takes precautions to avoid large step sizes or increasing residuals, [20]. This method requires the Jacobian of Eq. (10), which is computed easily in an exact form, and initial estimates of XP_1 and YP_2. The authors have used estimates that assume that the control points P_1 and P_2 will lay at approximately 25% and 50% of LWL. These assumptions produce realistic solutions for cases that the authors have tested.

Once XP_1 and YP_2 have been obtained, Eq. (2)–(5) give the remaining coordinates of the control points, and the constrained SAC is finally defined. Figure 3 presents SACs for various ship hulls defined with the presented method.

2.2. Constrained Definition of the Waterline

The method continues by defining the waterline of the ship hull (Figure 4), which is the intersection of the hull and the waterplane where the ship will float at rest. This curve confines a given area (waterplane area or AWP) and a centre of gravity (longitudinal centre of the waterplane or LCF), and covers the LWL.

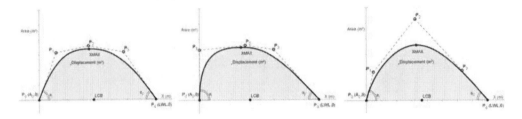

Figure 3. Examples of different Sectional Area Curves.

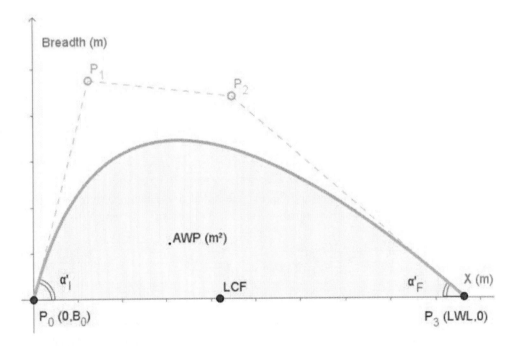

Figure 4. Parameters of the waterline.

As for the SAC, values and angles at the ends, α'_I and α'_F, set the final equations and can be used to move the area of the curve towards the ends while maintaining the total area. However, unlike the SAC, the position of the maximum is not set.

The waterline can not present a pronounced maximum, see Figure 5, and the variations of the maximum are not around the midpoint of the ship, as in the case of the SAC. Thus, the authors simplified the expression of the waterline using a cubic Bézier curve, Eq. (11):

$$\mathbf{WL}(t)=\sum_{j=0}^{3} B_j^3(t)\cdot\mathbf{P}_j \tag{11}$$

Figure 4 demonstrates that the unknowns are the positions of the control points $P_1(XP_1,YP_1)$ and $P_2(XP_2,YP_2)$ because the ends are fixed by the constraints. As in the case of the SAC, the tangent angles will produce two equations:

$$YP_1 = YP_0 + tg(\alpha'_I)\cdot(XP_1 - XP_0) \tag{12}$$

$$YP_2 = YP_3 + tg(\alpha'_F)\cdot(XP_3 - XP_2) \tag{13}$$

As in the previous section, the enclosed area of the WL is obtained by integration in a way similar to Eq.(6) after reducing the number of control points by one, yielding:

$$[\Phi]=\begin{bmatrix} -1 & \dfrac{3}{5} & \dfrac{3}{10} & \dfrac{1}{10} \\ -\dfrac{3}{5} & 0 & \dfrac{3}{10} & \dfrac{3}{10} \\ -\dfrac{3}{10} & -\dfrac{3}{10} & 0 & \dfrac{3}{5} \\ -\dfrac{1}{10} & \dfrac{3}{10} & -\dfrac{3}{5} & 1 \end{bmatrix} \quad AWP = [X]'\cdot[\Phi]\cdot[Y] \tag{14}$$

The abscissa of the centre of gravity of the waterline, LCF, can be computed by solving the integral (8) with just four control points. Using the same notation as in Eq. (8), the following matrix expressions are obtained:

$$[\Omega]=\begin{bmatrix} -\dfrac{1}{3} & \dfrac{3}{56} & \dfrac{3}{140} & \dfrac{1}{168} \\ -\dfrac{1}{8} & 0 & \dfrac{9}{280} & \dfrac{1}{28} \\ -\dfrac{1}{28} & -\dfrac{9}{280} & 0 & \dfrac{1}{8} \\ -\dfrac{1}{168} & -\dfrac{3}{140} & -\dfrac{3}{56} & \dfrac{1}{3} \end{bmatrix} \quad [\Psi]=\begin{pmatrix} \dfrac{1}{56} & \dfrac{1}{70} & \dfrac{1}{168} & \dfrac{3}{56} & \dfrac{1}{28} & \dfrac{1}{8} \\ \dfrac{3}{56} & \dfrac{3}{140} & \dfrac{1}{280} & \dfrac{9}{280} & 0 & \dfrac{3}{56} \\ \dfrac{3}{56} & 0 & -\dfrac{1}{280} & -\dfrac{9}{280} & -\dfrac{3}{140} & -\dfrac{3}{56} \\ \dfrac{1}{8} & \dfrac{1}{28} & -\dfrac{1}{168} & -\dfrac{3}{56} & -\dfrac{1}{70} & \dfrac{1}{56} \end{pmatrix}$$

$$AWP\cdot LCF = [Y]'\cdot[\Omega]\cdot\left[X^2\right]+[Y]'\cdot[\Psi]\cdot\left[\hat{X}\right] \tag{15}$$

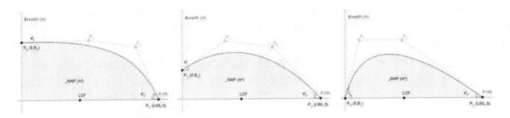

Figure 5. Examples of different waterlines.

As in the case of SAC, this equation is a function of the coordinates of the control points and of their powers. Equations (12)–(15) form a nonlinear system of four equations with four unknowns. As previously mentioned, a direct approach to this system can produce unrealistic solutions, and adequate initial estimates for the four unknowns are not immediately apparent.

The system can be simplified by substituting YP_1 and YP_2 from Eq. (12) and (13) into equations (14) and (15), yielding:

$$a'\cdot XP_1 + b'\cdot XP_2 + c'\cdot XP_1\cdot XP_2 + d' = AWP$$
$$f'\cdot XP_1^2 + g'\cdot XP_2^2 + h'\cdot XP_1^2\cdot XP_2 + j'\cdot XP_1\cdot XP_2^2 + k'\cdot XP_1\cdot XP_2 + l'\cdot XP_1 + m'\cdot XP_2 + n' = AWP\cdot XCF \quad (16)$$

This equivalent system is a function of XP_1 and XP_2 because a', b', ... n' are constants depending on the constraints. As in the SAC, this system is solved with a Powell hybrid algorithm, considering initial estimates for XP_1 and XP_2 of 25% and 75% of LWL, respectively. When the solutions have been computed, the remaining parameters that define the waterline are obtained by substitution into Eqs. (12) and (13), producing the final waterline. Figure 5 presents some examples of waterlines created with the presented method.

2.3. Constrained Definition of Round Bilge Hull Sections

All of the lines for the round bilge hull in Figure 1. have now been defined with the exception of the stations. A review of the constraints of Table 1 shows that these curves must be based on the SAC and waterline that have already been designed, enclosing a given area for the ship's draft T according to the SAC and with a given abscissa agreeing with the waterline, and will include the tangent angles at the centre and deck lines.

A direct B-spline approach to these stations does not produce realistic results. Considering the underwater part of the hull, any station has to enclose an area with the Y axis starting at a given angle and finishing at a given point at the waterline. All these constraints can be obtained with a second order Bézier curve, but the number of different ship hulls that can be obtained with such stations is limited. For example, ship hulls with an initial straight section cannot be reproduced. A higher order curve or a B-spline would require additional constraints such as curvature, which may not have a clear physical meaning for a naval architect and are difficult to calculate at the initial stage of the design.

Following Figure 6, where "n" indicates the index of the station, the proposed method will use the analytic expression $z = f(y)$ for the underwater part of the ship that starts at a point on the centreline (C_0) with a given angle (β_n), encloses an area ($S_n/2$) and interpolates the waterline point (C_1).

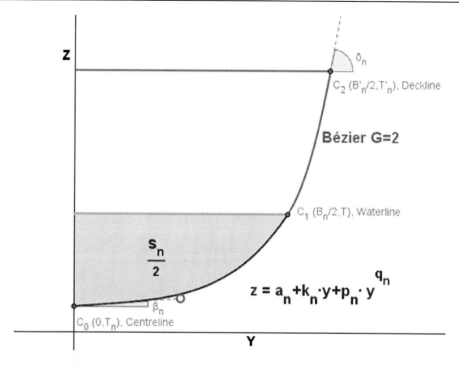

Figure 6. Definition of a station of index "n" for a round bilge hull.

The portion of the ship between the waterline (C₁) and the deckline (C₂) will be obtained with a second order Bézier curve because both the ends and tangent angles of the curve are known. The analytic expression (17) follows that of Jorde [21] and is the sum of a straight line and an exponent function. This is appropriate for ship hulls with sections that can start straight for hydrodynamic (dynamic lift) or construction (ease of development) purposes.

$$z = a_n + k_n \cdot y + p_n \cdot y^{q_n} \tag{17}$$

Integration of (17) indicates the values of the constants:

$$k_n = tg(\beta_n)$$
$$q_n = (T_n - k_n \cdot \frac{B_n}{2}) \cdot \frac{B_n}{2} \cdot \left[T_n \cdot \frac{B_n}{2} - \frac{S_n}{2} - \frac{k_n}{2} \cdot (\frac{B_n}{2})^2 \right]^{-1} - 1$$
$$p_n = (T_n - k_n \cdot \frac{B_n}{2}) / (\frac{B_n}{2})^{q_n}$$

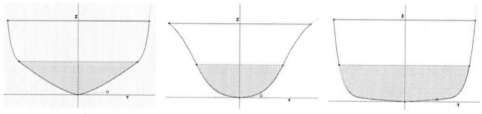

Figure 7. Examples of ship stations.

The above-water part of the ship is computed with a second order Bézier curve. The control points of this curve are C_1, C_2, and the intersection point of the tangent lines at these points. The angle at C_2 is a constraint of the method and the angle at C_1 can be computed from the first derivative of Eq. (17).

Figure 7 shows different sections generated with this method. The inclusion of the angles enables the designer to include concave/convex sections in the design. The convex configuration is used for the bow section of fast ships to deflect the spray of water away from the hull.

A CAD-compatible constrained B-Spline representation will be generated from a number of discrete points obtained from the analytic representation shown. This is detailed in the next section.

2.4. Constrained Approximation of a Set of Data Points

The next step of the method transforms points from the curves defined in the previous section into B-splines, considering together points from the under and above-water part of every station. This method will be also used in the definition of the hard-chine hull 3D lines. A large number of data points from the analytic curves does not suggest the use of an interpolating B-spline. Instead, an approximating curve is needed. For every station of index "d", the B-spline $c_d(u)$ will not pass through the data points exactly but will pass close enough to the points to capture the inherent shape. This is the well-known Least Squares (LS) approximation [22].

In this problem, $np+1$ data points Q_0, ...Q_{np} from the analytic curves will be approximated by a B-spline of pth degree, with $N+1$ control points P_0, ... P_N, $N < np$ that are unknown and are obtained as the final result of the calculations. These points will adopt constrained angles at the ends of the stations, angles β and δ in Figure 6, which should be included in the definition of the LS problem because they are constraints of the method.

The general LS problem is described by the over-determined set of $np+1+2$ equations with $N+1$ unknown variables:

$$B_0^p(t_0){\cdot}P_0 + B_1^p(t_0){\cdot}P_1 + ... + B_N^p(t_0){\cdot}P_N = Q_0$$

$$B_0^p(t_1){\cdot}P_0 + B_1^p(t_1){\cdot}P_1 + ... + B_N^p(t_1){\cdot}P_N = Q_1$$

$$\vdots \qquad \vdots \qquad \vdots \qquad \vdots$$

$$B_0^p(t_{np}){\cdot}P_0 + B_1^p(t_{np}){\cdot}P_1 + ... + B_N^p(t_{np}){\cdot}P_N = Q_{np} \tag{18}$$

$$B_0'^p(t_0){\cdot}P_0 + B_1'^p(t_0){\cdot}P_1 + ... + B_N'^p(t_0){\cdot}P_N = tg(\beta)$$

$$B_0'^p(t_{np}){\cdot}P_0 + B_1'^p(t_{np}){\cdot}P_1 + ... + B_N'^p(t_{np}){\cdot}P_N = tg(\delta)$$

Where B_i^p corresponds to the ith basis function of a pth degree B-spline that is calculated using de Boor's algorithm considering a uniform knot vector and t_j ($j=(0,n)$) represents the parameters associated to the data points. Matrix expressions are convenient to solve the problem:

$$\begin{bmatrix} B_0^P(t_0) & B_1^P(t_0) & \cdots & B_N^P(t_0) \\ B_0^P(t_1) & B_1^P(t_1) & \cdots & B_N^P(t_1) \\ \vdots & \vdots & \vdots & \vdots \\ B_0^P(t_{np}) & B_1^P(t_{np}) & \cdots & B_N^P(t_{np}) \\ B_0'^P(t_0) & B_1'^P(t_0) & \cdots & B_N'^P(t_0) \\ B_0'^P(t_{np}) & B_1'^P(t_{np}) & \cdots & B_N'^P(t_{np}) \end{bmatrix} \cdot \begin{bmatrix} P_0 \\ P_1 \\ \vdots \\ P_N \end{bmatrix} = \begin{bmatrix} Q_0 \\ Q_1 \\ \vdots \\ Q_{np} \\ tg(\beta) \\ tg(\delta) \end{bmatrix}; [M] \cdot [P] = [Q] \Rightarrow [M]^T \cdot [M] \cdot [P] = [M]^T \cdot [Q]$$

$$(19)$$

This system of equations is solved by multiplying both sides of Eq. (19) by $[M]^T$, which creates a determined (N+1) by (N+1) linear system. This type of system can be poorly conditioned, especially if a large number of control points is used. A conventional technique should not be used to solve this ill-conditioned system. Instead, single-value decomposition of $[M]^T \cdot [M]$ and a later back-substitution process is performed. The solutions of this system are the control points of the best B-Spline fitting. Approaching this problem with a standard parameterisation such as centripetal or chord-length is correct but does not consider the effect of the distance of the data points to the B-spline. In this method, we adopt a parameterisation based on a minimum distance. These parameterizations were first introduced by Hoschek [23], but the way to obtain the minimum distance is different. The process is iterative and is described by the following three steps:

1. The method starts with a centripetal parameterisation of the Q_i points and system (18) is solved. This produces a starting curve of the iterative process only for the first loop.
2. For each Q_i, the minimum distance to the B-Spline is calculated. This is done by dividing the B-Spline $c_d(u)$ in Bézier curves $b_j(u_L)$ ($j=1,N-p$) of the pth degree and computing the minimum distance to the corresponding Bézier piece, leading to a solution to Eq. (20).

$$(\mathbf{Q}_i - \mathbf{b}_j(u_L)) \cdot (\mathbf{b}'_j(u_L)) = 0 \qquad\qquad (20)$$

This equation is solved in the local domain of the Bézier curves. Because the equation is a polynomial equation, $u_L \in [0,1]$ and specific algorithms for this type of equation can be used. These algorithms do not require an initial guess, which would be required if a Newtonian method were used in the B-Spline domain. The current method uses a Jenkins-Traub 3-stage algorithm, [24]. The valid solution will be a non-complex solution of $u_L \in [0,1]$. Once the solution has been found, the local u_L for the Bézier domain is easily converted into its global value t_i in the B-spline domain. This t_i value is the parameter associated with the point Q_i when solving system (18).

1. After obtaining the ti ($i=1,np$) values, the distance di $= (Qi - c_d(ti))$ is computed, which is the Euclidean distance between Q_i and the B-Spline. This distance is used to check the shape requirement. If the maximum distance di ($i=1,n$) is above a given tolerance, steps 2 and 3 are repeated until an acceptable maximum distance is achieved. More specifically, the quality of the obtained curve is measured using the tolerance constraint and the shape of the B-Spline is amended using parameterisation (20).

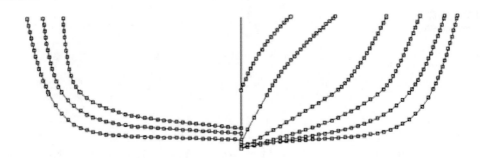

Figure 8. Example of the constrained fitting.

If the tolerance is not obtained in fewer than 50 iterations, then the number of control points $N+1$ has to be increased. The increment of the degree p in this procedure can also reduce the maximum distance. However, increasing the number of control points has a more substantial effect and a higher degree raises the complexity and the computational time. When the fitting is finished for all the stations, a set of q B-splines with the form of Eq. (37) are obtained. Figure 8 presents an example of the described fitting. This particular example shows a cubic B-spline fitting where all the curves have 9 control points and have been obtained after 30 iterations. The maximum distance from the data points to the curves is less than 2 mm, and the dimensions of this ship are 6690 mm x 2000 mm. The tolerance can be improved by increasing the number of control points as needed to satisfy construction requirements. The last step of the method is to create a B-spline surface that leans on the stations previously defined. This is also common for the hard-chine hull definition, so it will be explaned later in section 0.

3. DESCRIPTION OF THE METHOD FOR A HARD-CHINE HULL

The geometry of a single hard-chine hull, also called planing hull, is well represented by the lines in Figure 1 down. The way of designing a hard-chine hull is very different from the method described for a round bilge hull of the previous sections, because the hard chine is more geometrical in nature, and a SAC and WL curves are not used in the design process. A hard-chine hull has a flat transom stern and can be modeled by decomposing the surface into boundaries that will be constrained by the design parameters. These numerical parameters include position and slope. In the case of the chine, which is the most significant curve of a planing hull design, the enclosed area (Ac) and centroid (X_C) are also included in the numerical parameters. The boundary curves are the keel or center line (CL), the chine line and the sheer line. The objective is to create B-spline surfaces to represent a ship hull based on the constraints of Table 2, with the graphical meaning depicted in Figure 9. The parameters in Table 2 have the graphical meaning depicted in Figure 9, where the vectors indicate the angle between the X-axis and the arrow line. The selected parameters used to define these curves have meaning for the ship designer and can be related to further aspects of ship design. The curve definition has been simplified by considering that the hull has a flat vertical transom and that the aftermost point of all the curves has a zero abscissa as in Figure 9.

Table 2. Parameters of the method (see also Figure 9)

Name	Description	Location
	Length	
Ls	Abscissa of the forward-most point of the sheer line	Center, Sheer
L0	Abscissa where the forefoot is tangent to the keel line	Center
Lx	Abscissa of the sheer's maximum breadth	Sheer (plan)
Lc	Abscissa of the forward-most point of the chine	Center, Chine
Xc	Abscissa of the centroid of Ac	Chine (plan)
X_{C1}	Abscissa of an intermediate point of the chine	Chine (profile)
	Width	
Bs	Sheer's half-breadth at the transom	Sheer (plan)
Bx	Ordinate of the sheer's maximum half-breadth	Sheer (plan)
Bc	Chine's half-breadth at the transom	Chine (plan)
Sp	Width of the spray rail at the transom	3D
	Height	
Hs	Height of the foremost point of the sheer	Center, Sheer (profile)
Hc	Height of the foremost point of the chine	Center, Chine (profile)
r	Rocker at the transom	Center
hs	Sheer's height at the transom	Sheer (profile)
hc	Chine's height at the transom	Chine (profile)
Z_{C1}	Height of X_{C1}, normally the draft of the ship.	Chine (profile)
	Angles	
α_K	Angle at the stem	Center
α_S	Angle at the foremost point of the sheer in plan view	Sheer (plan)
β'_S	Angle at the transom of the sheer in lateral view	Sheer (profile)
α'_S	Angle at the foremost point of the sheer in lateral view	Sheer(profile)
α_C	Angle at the foremost point of the chine in plan view	Chine (plan)
β_C	Angle at the transom point of the chine in plan view	Chine (plan)
α'_C	Chine's angle at the foremost point in profile view	Chine (profile)
β'_C	Chine's angle at the transom point in profile view	Chine (profile)
	Areas	
$2 \cdot Ac$	Enclosed area between the chine and the X-axis in plan view	Chine (plan)

In the case of a non-vertical transom, the definition can be easily reconfigured to consider a non-zero abscissa for the aforementioned points, which will be a function of the transom angle. A very important design feature, the transom dead-rise angle, Ω, is deducted from hr, hc, and Bc, $\Omega = \text{Arc} \tan(\frac{hc - hr}{Bc}) \cdot \frac{180}{\pi}$ in degrees. The overall maximum dimensions of the length and breadth of the hull are Ls and $2 \cdot Bx$, respectively. The method allows the definition of a spray rail of a given width, Sp, along the chine, as will be explained in 0.4. The method splits up the stations into two different domains: below and above the chine. The method controls the concavity /convexity of these curves using a parameter that controls the maximum deviation of every station piece from a straight segment, as detailed in 0.7. In the following sections, background figures corresponding to real ship designs, have been used to show how the B-spline curves of the method produce realistic results in the figures of the following sections.

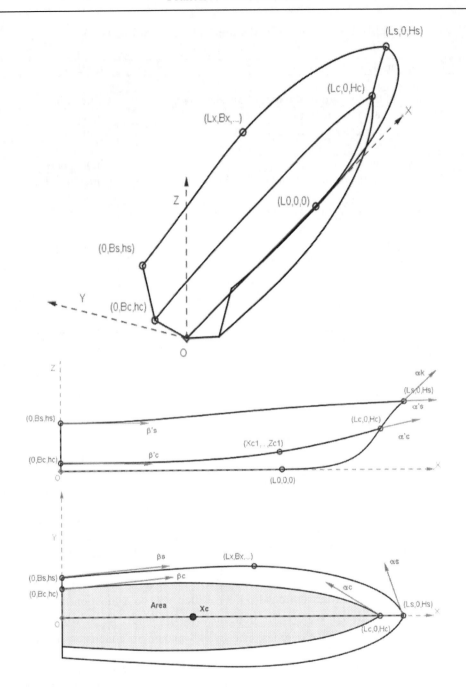

Figure 9. Graphical representation of the parameters.

3.1. Definition of the Center Line

In a hard-chine hull, the center or keel line is usually without appendages or a skeg and runs in a straight line forward to the transom (O). This line curves up toward the stem at the start of the forefoot $K_0(L_0, 0)$ and arrives at the stem at $K_2(Ls, Hs)$ with a given angle ($α_K$). The center line contains the chine's fore end $K_1(Lc, Hc)$.

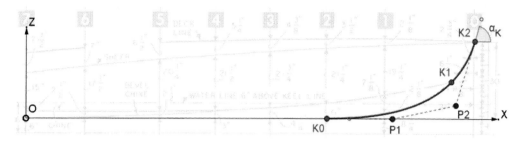

Figure 10. Center line definition.

With these assessments, the geometric problem consists of finding a B-spline starting at K_0 with a zero angle and arriving at K_2 with an angle α_K and then interpolating a point K_1. This can be obtained with a cubic B-spline of 4 control points with the form of:

$$c(u) = B_0^3 \cdot K_0 + B_1^3 \cdot P_1 + B_2^3 \cdot P_2 + B_3^3 \cdot K_2$$

A particularization of the basis functions B_i^j and their derivatives for cubic and quadratic B-splines of 4 and 3 control points and a uniform knot vector are shown in Table 3. The curves of this section and following ones use these functions in their definitions. The unknown values will be the coordinates of $P_1(XP_1, ZP_1)$ and $P_2(XP_2, ZP_2)$ because the curve will contain the end points K_0 and K_2, and they will be obtained after imposing the constraints. The first two constraints are related to the tangent angles at the ends of the curve: $c_z'(0) = 0$ and $c_z'(1) = tg(\alpha_K)$. The third constraint indicates that the curve interpolates the point K_1 for a certain parameter u^*: $c(u^*) = K_1$. The value of this parameter is selected according to the centripetal parameterization, which depends on both the Euclidean distance (Dist( )) of K_1 to the ends of the curve and an exponent k:

$$u^* = \frac{\text{Dist}(K_0, K_1)^k}{\text{Dist}(K_0, K_1)^k + \text{Dist}(K_1, K_2)^k} \qquad (21)$$

The value $k=1$ produces a chord-length parameterization and gives good results for the center line. The constraints $c'(0) = 0$, $c'(1) = tg(\alpha_K)$ and $c(u^*) = K_1$ produce a linear system of four equations with the following matrix form:

$$\begin{bmatrix} 0 & 1 & 0 & 0 \\ 0 & 0 & -tg(\alpha_K) & 1 \\ B_1^3(u^*) & 0 & B_2^3(u^*) & 0 \\ 0 & B_1^3(u^*) & 0 & B_2^3(u^*) \end{bmatrix} \begin{bmatrix} XP_1 \\ ZP_1 \\ XP_2 \\ ZP_2 \end{bmatrix} = \begin{bmatrix} 0 \\ Hs - tg(\alpha_K) \cdot Ls \\ Lc - B_0^3(u^*) \cdot L_0 - B_3^3(u^*) \cdot Ls \\ Hc - B_3^3(u^*) \cdot Hs \end{bmatrix} \qquad (22)$$

The solution of this system of equations, which can be obtained with, for example, the Gauss-Jordan method, produces the final form of the center line. Notice how a plumb bow (Figure 11) can be obtained, thereby reducing the forefoot radius by increasing the value of α_K and moving K_2 backward closer to K_1.

Table 3. Basis and derivative functions of the B-splines for the presented method

j	B^3_j	B'^3_j	B^2_j	B'^2_j
0	$1-3 \cdot u+3 \cdot u^2-u^3$	$-3+6 \cdot u-3 \cdot u^2$	$(1-u)^2$	$-2 \cdot (1-u)$
1	$3 \cdot u-6 \cdot u^2+3 \cdot u^3$	$3-12 \cdot u+9 \cdot u^2$	$2 \cdot u \cdot (1-u)$	$2-4 \cdot u$
2	$3 \cdot u^2-3 \cdot u^3$	$6 \cdot u-9 \cdot u^2$	u^2	$2u$
3	u^3	$3 \cdot u^2$	0	0

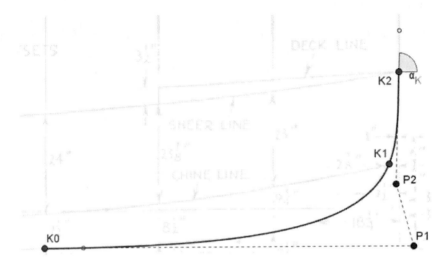

Figure 11. Plumb bow.

The shape of the forefoot can be altered while maintaining the rest of the parameters by considering different values for k in Eq. (21) as depicted in Figure 12. This can be used to reduce or to increase the dead-rise angle of the forefoot sections.

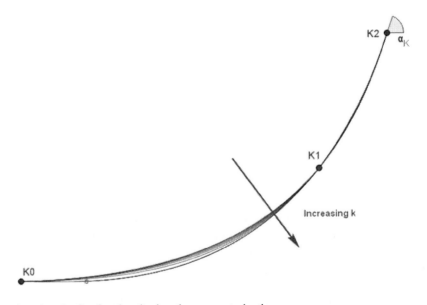

Figure 12. Changing the forefoot by altering the parameterization.

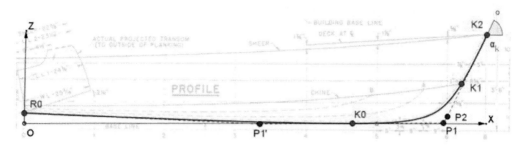

Figure 13. Rocker in the afterbody.

Some designs may include a certain amount of rocker *hr* in the afterbody, or even some hook (negative *hr*), but the latter is not common in contemporary designs. This means that the center line does not run straight from the transom and that the part of the center line abaft of K$_0$ has to be modeled in a different way. Realistic results can be obtained with a 2nd degree B-spline, $n=2$, which is constructed as depicted in Figure 13 with only 3 control points: R$_0$(0,*hr*), P'$_1$ and K$_0$, where P'$_1$ is the symmetric point of P$_1$ with respect to K$_0$. This insures that the aft part of the center line and the fore part, calculated in (22), are of C^1 class continuity at their joint point K$_0$. The use of different pieces to define a curve is not a problem for the proposed method and increases its flexibility because the curves can be converted into a single B-spline by considering different points on both curves. This will be explained in Section 0.6. This technique is also used to produce the 3D curves of the chine and sheer lines from their 2D orthogonal projections. The forefoot contour is related to the directional stability in smooth and cross seas; cross seas tend to throw the bow around. A deep forefoot with a high value of $α_K$ produces a narrow bow and increases the dead-rise angle of the bow sections. Styling is also a variable to be considered. There is a trend toward the use of plumb bows in new designs, especially in Europe, although it is not recommended from a hydrodynamic point of view for a planing boat. A clipper bow that has an "S" shape above the waterline can be achieved by reducing $α_K$. This bow works well with a short bowsprit or anchor handling platform. The use of a rocker is common in semi-displacement hulls designed to go more through the water than planing on top, because the aft buttocks are not straight. The use of hook is not common today, and a better effect can be obtained with the use of adjustable trim tabs.

3.2. Definition of the Sheer Line in the Plan View

In a plan view, the sheer line runs from the transom towards the stem and may have a maximum if the breadth at the transom is lower than the maximum breadth, or the maximum breadth is constant aftwards, which is a common design trend. In the first case (Figure 14), the sheer line starts at the transom, at a point S$_0$(0, *Bs*), has a maximum at S$_X$(*Lx*, *Bx*), and arrives at the point S$_2$(*Ls*, 0) with a given angle $α_S$. The restraints of this line, s$_p$(u), will be s$_p$(0) = S$_0$, s$_p$(1) = S$_2$, and s$_p$(u*) = S$_x$, together with s'$_{py}$(u*) = 0 and s'$_p$(1) = tg($α_S$). Notice that the tangent angle at the transom is not imposed. These requirements are achieved with a cubic B-spline with four control points:

$$s_p(u) = B_0^3 \cdot S_0 + B_1^3 \cdot P_1 + B_2^3 \cdot P_2 + B_3^3 \cdot S_2$$

The unknown values are the coordinates of $P_1(XP_1, YP_1)$ and $P_2(XP_2, YP_2)$ because the curve contains the end points S_0 and S_2. These values will be obtained after imposing the constraints. The parameter u^* is selected in the same manner as the center line, Eq. (21), with the chord length parameterization. The constraints $s_p(0) = S_0$, $s_p(1) = S_2$, and $s_p(u^*) = S_x$ together with $s'_{py}(u^*) = 0$ and $s'_p(1) = tg(\alpha_S)$ produce a linear system of four equations with the following matrix form:

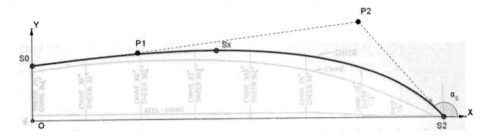

Figure 14. Sheer line definition in plan view.

$$\begin{bmatrix} 0 & B'^3_1(u^*) & 0 & B'^3_2(u^*) \\ 0 & 0 & -tg(\alpha_S) & 1 \\ B^3_1(u^*) & 0 & B^3_2(u^*) & 0 \\ 0 & B^3_1(u^*) & 0 & B^3_2(u^*) \end{bmatrix} \cdot \begin{bmatrix} XP_1 \\ YP_1 \\ XP_2 \\ YP_2 \end{bmatrix} = \begin{bmatrix} -B'^3_0(u^*) \cdot Bs \\ -tg(\alpha_S) \cdot Ls \\ Lx - B^3_3(u^*) \cdot Ls \\ Bx - B_0(u^*) \cdot Bs \end{bmatrix} \quad (23)$$

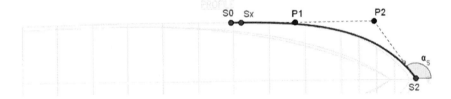

Figure 15. Sheer line with constant width.

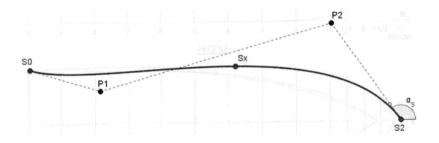

Figure 16. Non-realistic sheer.

Where B^3_j and B'^3_j are obtained from Table 3 and u^* is calculated following Eq. (21) particularized for S_0, S_x and S_2. The solution of this system produces the control points P_1 and P_2, and the plan view of the sheer line is completely defined. If the sheer maintains its maximum breadth aft, then expression (23) is still valid without considering S_0 at the transom, as long as it is close enough to S_x, as depicted in Figure 15.

In this case, the curve is correctly defined between S_x ($u=u^*$) and S_2 ($u=1$). If S_0 is placed at the transom, as in Figure 16, then expression (23) produces results that are mathematically correct but are not realistic for the part of the curve between S_0 ($u=0$) and S_x ($u=u^*$). This is because $n+1$ aligned control points are needed to define a straight portion of an nth degree B-spline, which is not possible in this particular case.

The definition of a sheer line with a constant width sheer, according Figure 15, is not a problem for the method because the 3D curves will be obtained based on points from the 2D curves. Therefore the straight portion, which is not considered, will be modeled using a straight segment between the transom and S_x. Another way to vary the aspect of the curve is to consider different values of k, apart from $k = 1$, in the parameterization of Eq. (21).

The definition of the sheer line is related to the interior volume distribution above the water, which is linked to the angle α_S; namely, a higher angle increases the useful volume at the fore part of the ship and vice versa. This curve is also related with seakeeping because a reduction in α_S reduces the angle of the bow sections above the water, producing a wider bow. This can produce pounding when advancing in high waves, but it will produce a dry ship in moderate sea states because the water is deflected more pronouncedly than with narrower sections.

3.3. Definition of the Sheer Line in the Lateral View

The sheer line in a profile view, runs from the transom towards the stem in correspondence with its projection in the plan view. Normally, the sheer line will present neither a maximum nor minimum in a planing hull. As a high-speed boat starts onto a plane, it trims up at the bow. With a conventional sheer line, the view over the bow becomes obstructed and, at a critical moment, the view from the helm can be completely blocked. Therefore, the sheer line in the lateral view will start at a given height $S'_0(0, hs)$ at the transom with a given angle β'_S and will arrive at $S'_2(Ls, Hs)$ at the stem with a given angle α'_S, as described in Figure 17. The point S'_2 of the sheer line is equivalent to K_2 of the center line of Figure 10.

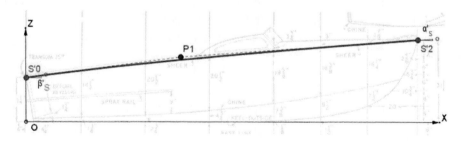

Figure 17. Sheer definition in the lateral view.

The restraints of the sheer line in the profile view, $s_L(u)$, will be $s_L(0) = S'_0$, $s_L(1) = S'_2$, $s'_L(0)=tg(\beta'_S)$ and $s'_L(1)=tg(\alpha'_S)$. These requirements are met with a 2nd degree B-spline with three control points of the form:

$$s_L(u) = B_0^2 \cdot S'_0 + B_1^2 \cdot P_1 + B_2^2 \cdot S'_2$$

The unknown values are the coordinates of $P_1(XP_1, ZP_1)$ because the curve will contain the end points S'_0 and S'_2. These values will be obtained after imposing the constraints. The following linear system of equations has to be solved, as in the previous sections:

$$\begin{bmatrix} tg(\beta'_s) & -1 \\ -tg(\alpha'_s) & 1 \end{bmatrix} \cdot \begin{bmatrix} XP_1 \\ ZP_1 \end{bmatrix} = \begin{bmatrix} -hs \\ Hs - tg(\alpha'_s) \cdot Ls \end{bmatrix} \tag{24}$$

In modern designs, the sheer line in the profile view sometimes runs straight. This flat sheer line cannot be modeled with (24) because the system of equations has no real solution when $\beta'_s = \alpha'_s$. This can be solved using the properties of the B-spline; aligning the degree+1 control points produces a straight part of the curve.

Therefore, considering $P_1(XP_1, ZP_1)$ as the middle point of S'_0 and S'_2, produces a valid solution (Figure 18). The lateral view of the sheer line is related to the aesthetics and is the most visual line of a ship.

The height of S'_2 corresponds to good visibility from the helm station, and this value is also a function of its placement in the general arrangement. If the helm is placed forward, the freeboard Hs can be raised to achieve drier conditions in waves; however, a high Hs in a small boat tends to be non-aesthetic. A common solution is to reduce the freeboard and raise the helm position.

3.4. Definition of the Chine Line in the Plan View

The chine is the most significant line of a planing hull because it controls the attitude of the craft to go onto a plane. The chine runs from the transom towards the stem (Figure 19), starting at a point, $C_0(0, Bc)$, with a given angle, β_c, and arriving at the stem at $C_2(Lc, 0)$ with an angle, α_c.

In this case, the method will impose a confined area (Ac) and the longitudinal position of the center of the enclosed area (Xc). According to Blount [15] and Saunders [25], the area and its centroid have a direct influence on the planing behavior and also on dynamic instabilities of the design. If the hull has a spray rail in the chine (Figure 29), only the inner part of the chine is considered. The method used to include a spray rail will be explained in Section 0.6 when 3D curves are created.

Mathematically, the above conditions can be imposed with a cubic B-spline with four control points, in a similar way as the waterplane of a round bilge hull of Section 0.2:

$$c_p(u) = B_0^3 \cdot C_0 + B_1^3 \cdot P_1 + B_2^3 \cdot P_2 + B_3^3 \cdot C_2$$

The unknown values are the coordinates of $P_1(XP_1, YP_1)$ and $P_2(XP_2, YP_2)$ because the curve will contain the extreme points C_0 and C_2, and P_1 and P_2 will be obtained after imposing the two angle constraints, the enclosed area and its centroid. An important difference between the chine and previously defined lines, is that imposing the area and its centroid generates a non-linear problem.

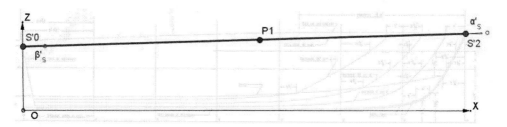

Figure 18. Flat sheer definition.

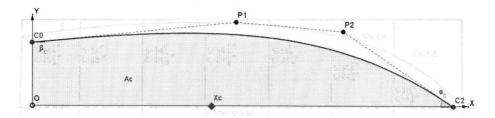

Figure 19. Chine definition in the plan view.

The angle constraints are introduced as in the previous curves by considering the properties of the control polygon of a B-spline at its ends, which produces the following equations:

$$YP_1 = YC_0 + tg(\beta_c) \cdot (XP_1) \tag{25}$$

$$YP_2 = YC_2 + tg(\alpha_c) \cdot (Lc - XP_2) \tag{26}$$

The most difficult part of the problem is the inclusion of the enclosed area and its center of gravity into the definition of the problem. Although the area of the closed B-spline can be easily computed with Greens' theorems, in this case, the area is enclosed by the curve and the X-axis. This can be solved by integration in the parametric domain:

$$Ac = \int dA = \int_0^{Lc} Y(t)\,dX = \int_0^1 Y(t)X'(t)\,dt \tag{27}$$

For this particular case, a cubic B-spline with 4 control points, *X'(t)* and *Y(t)* are cubic polynomials computed with the cubic basis functions of the B-spline and their derivatives (Table 3). The result of the integral of Eq. (27) can be expressed in matrix form as in Eq. (14)

$$[X] = \begin{pmatrix} 0 \\ XP_1 \\ XP_2 \\ Lc \end{pmatrix} \quad [Y] = \begin{pmatrix} Bc \\ YP_1 \\ YP_2 \\ 0 \end{pmatrix} \quad [\Phi] = \begin{pmatrix} \Phi_{00} & \cdots & \Phi_{03} \\ \vdots & \ddots & \vdots \\ \Phi_{30} & \cdots & \Phi_{33} \end{pmatrix} \quad \Phi_{ij} = \int_0^1 B_i^3(t) B_j'^3(t)\,dt$$

$$[\Phi] = \begin{bmatrix} -1 & \dfrac{3}{5} & \dfrac{3}{10} & \dfrac{1}{10} \\[2mm] -\dfrac{3}{5} & 0 & \dfrac{3}{10} & \dfrac{3}{10} \\[2mm] -\dfrac{3}{10} & -\dfrac{3}{10} & 0 & \dfrac{3}{5} \\[2mm] -\dfrac{1}{10} & -\dfrac{3}{10} & -\dfrac{3}{5} & 1 \end{bmatrix} \qquad Ac = [X]^t [\Phi] [Y] \tag{28}$$

The abscissa of the center of gravity of the chine's enclosed area, Xc, can be computed by solving the following integral:

$$Xc = \frac{\int X \, dA}{Ac} = \frac{\int_0^{Lc} X(t)\,Y(t)\,dX}{Ac} = \frac{\int_0^1 X(t)\,Y(t)\,X'(t)\,dt}{Ac} \tag{29}$$

As in the previous calculations of the enclosed area, expression (29) can be computed with the cubic basis functions and their derivatives and expressed in matrix form:

$$[X^2] = \begin{pmatrix} 0^2 \\ XP_1^2 \\ XP_2^2 \\ Lc^2 \end{pmatrix} \quad [\hat{X}] = \begin{pmatrix} Lc \cdot XP_2 \\ Lc \cdot XP_1 \\ Lc \cdot 0 \\ XP_2 \cdot XP_1 \\ XP_2 \cdot 0 \\ XP_1 \cdot 0 \end{pmatrix} \quad [\Omega] = \begin{pmatrix} \Omega_{00} & \cdots & \Omega_{03} \\ \vdots & \ddots & \vdots \\ \Omega_{30} & \cdots & \Omega_{33} \end{pmatrix} \quad \Omega_{ij} = \int_0^1 B_i^3(u)\,B_j^3(u)\,B_j'^3\,du$$

$$[\Psi] = \begin{pmatrix} \Psi_{32}^0 & \cdots & \Psi_{30}^0 & \Psi_{21}^0 & \Psi_{20}^0 & \Psi_{10}^0 \\ \vdots & \ddots & \vdots & \vdots & \vdots & \vdots \\ \Psi_{32}^4 & \cdots & \Psi_{30}^3 & \Psi_{21}^3 & \Psi_{20}^3 & \Psi_{10}^3 \end{pmatrix}$$

$$\Psi_{ij}^p = \int_0^1 B_p^3 \left[B_i^3(u)\,B_j'^3(u) + B_j^3(u)\,B_i'^3(u) \right] du$$

$$[\Omega] = \begin{bmatrix} -\dfrac{1}{3} & \dfrac{3}{56} & \dfrac{3}{140} & \dfrac{1}{168} \\[2mm] -\dfrac{1}{8} & 0 & \dfrac{9}{280} & \dfrac{1}{28} \\[2mm] -\dfrac{1}{28} & -\dfrac{9}{280} & 0 & \dfrac{1}{8} \\[2mm] -\dfrac{1}{168} & -\dfrac{3}{140} & -\dfrac{3}{56} & \dfrac{1}{3} \end{bmatrix} \qquad [\Psi] = \begin{pmatrix} \dfrac{1}{56} & \dfrac{1}{70} & \dfrac{1}{168} & \dfrac{3}{56} & \dfrac{1}{28} & \dfrac{1}{8} \\[2mm] \dfrac{3}{56} & \dfrac{3}{140} & \dfrac{1}{280} & \dfrac{9}{280} & 0 & -\dfrac{3}{56} \\[2mm] \dfrac{3}{56} & 0 & \dfrac{1}{280} & -\dfrac{9}{280} & \dfrac{3}{140} & -\dfrac{3}{56} \\[2mm] -\dfrac{1}{8} & -\dfrac{1}{28} & -\dfrac{1}{168} & -\dfrac{3}{56} & -\dfrac{1}{70} & -\dfrac{1}{56} \end{pmatrix}$$

$$Ac \cdot Xc = [Y]^t \cdot [\Omega] \cdot [X^2] + [Y]^t \cdot [\Psi] \cdot [\hat{X}] \tag{30}$$

Equation (31) is a function of the cross product of the control point coordinates and their squared and cubic powers. This yields four equations (25), (26), (28), and (30) with four unknowns, XP_1, XP_2, YP_1 and YP_2, which form a nonlinear system of equations that must be solved. A direct approach to the solution of such a system can be numerically difficult to obtain because the solution is very sensitive to initial estimates of the solutions. In addition, non-realistic results can be obtained because of the nonlinearity of Eqs. (28) and (30). Manipulation of the linear conditions of Eqs. (25), and (26) enables further simplifications, by substituting YP_1 and YP_2 into Eqs. (28) and (30), yielding:

$$\begin{aligned} & a'\cdot XP_1 + b'\cdot XP_2 + c'\cdot XP_1 \cdot XP_2 + d' = Ac \\ & f'\cdot XP_1^2 + g'\cdot XP_2^2 + h'\cdot XP_1^2 \cdot XP_2 + j'\cdot XP_1\cdot XP_2^2 + k'\cdot XP_1\cdot XP_2 + l'\cdot XP_1 + m'\cdot XP_2 + \\ & +n' = Ac\cdot Lc \end{aligned} \quad (31)$$

Where a', b', c',…n' are constants that depend on the initial numerical constraints. The solution of this nonlinear system can be calculated with a Powell hybrid algorithm as was done with the waterplane of a round bilge hull in Section 0.2. This method requires the Jacobian of Eq. (31), which is computed easily in an exact form, and initial estimates of XP_1 and XP_2. Good results are obtained if the estimates of the control points, P_1 and P_2, lie at approximately 50 and 75% of Lc. These assumptions produce realistic solutions in the cases for which they have been tested.

After XP_1 and XP_2 have been obtained, Eqs. (25), and (26) give the remaining coordinates of the control points, and the constrained chine in the plan view is finally defined. If the maximum breadth of the chine is constant, as in the case of prismatic hulls, the solution of (31) considering $\beta c = 0$, is mathematically valid. However, this solution is not practical in a similar manner as the sheer curve of Figure 15 and Figure 16. The reason is the same as that in the case of the sheer line; namely, more than four controls points are needed to reproduce a straight part of a cubic B-spline.

The mathematical model for the chine can be maintained if the constraints are corrected without considering the prismatic part of the chine, as depicted in Figure 20. In this case, the length of the prismatic body is deducted to $\mathbf{C_2}$ and Xc, and the area of the prismatic part is deduced to Ac; thus, system (31) is still valid with these assumptions.

As in the case of the sheer with a constant breadth, the straight part of the curve will be considered to obtain the final 3D definition of the curve.

The shape of the chine in the plan view is directly related to the planing behavior of the craft. The breadth at the transom is reduced to improve seakeeping because this reduces the wave forces in following seas and increases the dead-rise angle at the transom. As the design speed increases, Xc is moved aft to control the dynamic trim angle and to obtain a finer fore body. Remember that the chine's breadth (Bc) and height (hc) at the transom are interrelated with the dead-rise angle at the transom, Ω.

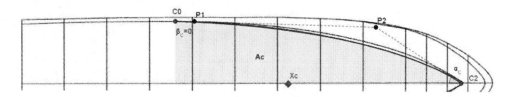

Figure 20. A chine of constant breadth.

3.5. Definition of the Chine Line in the Lateral View

The chine in the profile view, Figure 21, runs from the transom at a point C'$_0$(0, hc), to C'$_2$(Lc, Hc) in the center line. C'$_2$ is equivalent to K$_1$(Lc, Hc) in Figure 10. The line will start at the transom with a given angle, β'_c, and will arrive at the centerline with an angle α'_c, crossing through a point C'$_1$(X$_{C1}$, Z$_{C1}$) that is imposed by the designer. Mathematically, this problem is quite similar to the definition of the center line, although the angle at the aft point of the curve could be non-zero in this case. This curve is defined with a four control point cubic B-spline such as the following:

$$c_L(u) = B_0^3 \cdot C'_0 + B_1^3 \cdot P_1 + B_2^3 \cdot P_2 + B_3^3 \cdot C'_2$$

The unknown values are the coordinates of P$_1$(XP$_1$, ZP$_1$) and P$_2$(XP$_2$, ZP$_2$) because the curve will contain the extreme points C'$_0$ and C'$_2$. These values will be obtained after imposing the constraints. The first two constraints are related to the tangent angles at the ends of the curve: $c_L'(0) = \text{tg}(\beta'_c)$ and $c'_L(1) = \text{tg}(\alpha'_c)$. The last constraint indicates that the curve interpolates the point C$_1$ for a certain parameter u^*: $c_L(u^*) = C_1$, and u^* is calculated according to expression (21), considering $k = 1$. As in the case of the center line, the constraints produce a linear system of equations with the following matrix form:

$$\begin{bmatrix} -\text{tg}(\beta'_c) & 1 & 0 & 0 \\ 0 & 0 & -\text{tg}(\alpha'_c) & 1 \\ B_1^3(u^*) & 0 & B_2^3(u^*) & 0 \\ 0 & B_1^3(u^*) & 0 & B_2^3(u^*) \end{bmatrix} \begin{bmatrix} XP_1 \\ ZP_1 \\ XP_2 \\ ZP_2 \end{bmatrix} = \begin{bmatrix} h_c \\ Hc - \text{tg}(\alpha'_c) \cdot Lc \\ X_{C1} - B_3^3(u^*) \cdot Lc \\ Z_{C1} - B_0^3(u^*) \cdot hc - B_3^3(u^*) \cdot Hc \end{bmatrix} \quad (32)$$

The solution of this system produces control points P$_1$ and P$_2$, and the lateral view of the chine line is completely defined. The lateral projection of the chine largely determines the fore and aft shape of the buttocks, particularly in the after body. Straight buttocks in the after body wetted region reduce negative differential pressures in this area and prevent bottom suction and excessive trim by the stern. The angle β'_c will be nearly zero; therefore, the shape of the chine in the after body is also controlled by the position of the point C'$_1$, which can be taken as a reference at the point where the chine intersects the water plane. Therefore, the convexity of the chine, and in turn of the buttocks, is controlled both with a low value of β'_c and with the position of C'$_1$, which is normally forward of 50% of the ship's waterline length and at the draught's height, [25].

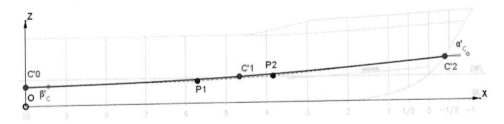

Figure 21. Definition of the chine in the profile view.

Parametric Design of Small Ships with the Use of NURBS Surfaces 35

A simpler definition of the chine line could be made with a three control point B-spline, avoiding the intermediate point C'_1, as in the case of the lateral projection of the sheer line. However, the presented curve with 4 control points provides better control over the curve's shape. The chine's depth in the aft body is determined by the transom dead-rise angle, which is of paramount importance for the hydrodynamic behavior of the design. This point should be slightly below the pretended draft to guarantee that the initial stability of the design will be adequate.

Regarding the chine shape in the fore body, the higher the chine, the better the seakeeping because of the increment of the dead-rise angle in the fore sections. However, this sacrifices the internal volume in this area, and excessive height reduces the buoyancy of the bow.

3.6. Generation of the 3D Curves

In the preceding sections, the orthographic projections of the main curves of the ship hull have been defined. A direct 3D approach to the definition of the lines would have been much more difficult and less realistic from the design point of view because these ships are designed based on 2D sketches of the general arrangement and their associated 2D curves. Nevertheless, naval architecture software works with a 3D definition of the hull. Therefore, the next part of the method is to create a 3D definition of the ship's main lines (upper part of Figure 9) based on the orthographic projections.

This step is carried out based on a minimum squared fitting of 3D data points Qi, obtained from 2D projections of the curves. These points are obtained by considering the intersection of every pair of curves (center line alone, sheer in the plan and profile views, chine in the plan and profile views) with a set of np +1 abscissas XQi, $i= 0$, np. The corresponding data points for every line, Qi (XQi, YQi, ZQi), $i = 0$, np, are obtained in this way.

These abscissas XQi, will lie in the interval $[0, Ls]$ for the center and sheer lines and in $[0, Lc]$ for the chine. It is not necessary to consider them to be uniformly spaced inside the intervals. More data points can be selected in the forward part of the curves, where the curvature is more pronounced than in the aft part, to produce a more accurate fitting. In addition, a different value of np can be used for the chine, sheer and center lines.

Considering a generic cubic curve of the method with four control points and with orthographic projections c(u) and c'(u), the 3D points are obtained through the following steps:

- $c(u) = (X(u), Y(u)) = B_0^3 \cdot P_0 + B_1^3 \cdot P_1 + B_2^3 \cdot P_2 + B_3^3 \cdot P_3; \quad P_i = (XP_i, YP_i)$

- Obtain the value of u that makes $c(u) = XQ_i$ by solving:

$$XP_0 + u \cdot (-3 \cdot XP_0 + 3 \cdot XP_1) + u^2 \cdot (3 \cdot XP_0 - 6 \cdot XP_1 + 3 \cdot XP_2) +$$
$$+u^3 \cdot (-XP_0 + 3 \cdot XP_1 - 3 \cdot XP_2 + P_3) = XQ_i \tag{33}$$

- Substitute u into $Y(u)$ to obtain $YQ_i = Y(u)$
- Repeat the procedure in $c'(u) = (X(u), Z(u))$ to obtain ZQ_i, and for $i = 0, np$

Equation (33) is obtained by expanding the equation of a cubic B-spline with 4 control points by considering the basis functions of Table 3. In the case of the center line, there is only its profile view. If a rocker value is used (Figure 13), the procedure must be repeated separately for the two curves that form the center line. If this line does not present rocker (Figure 10), the points on the straight segment between the origin and the point K_0 are easy to obtain.

In the case of the sheer line in the lateral view (Figure 17), the curve is modeled with just 3 control points instead of four. This produces a simplification in the procedure, and Eq. (33) is substituted by the following equation:

$$XP_0 + u \cdot (-2 \cdot XP_0 + 2 \cdot XP_1) + u^2 \cdot (XP_0 - 2 \cdot XP_1 + XP_2) = XQ_i \tag{34}$$

After $np + 1$ data points Q_i, have been obtained, the fitting begins. This fitting is produced according the methodology explained in Section 0.4 for the definition of the stations of a round-bilge hull. When the fitting is finished for all the curves, a set of 3 B-splines is obtained. Figure 22 presents an example of the described fitting. This particular example shows a cubic B-spline fitting of a ship of maximum dimensions Ls = 16.94 m, Bx = 2.42 m and Hs = 3.27 m. The sheer line has been modeled with 10 control points, with a maximum tolerance of 0.008 m and a medium tolerance of 0.002 m. The chine has been modeled with 7 control points, and the tolerances are max. = 0.007 m and med. = 0.004 m. The center line has 19 control points and tolerances of max. = 0.009 m, med.= 0.001 m. The curves need more control points to achieve a given tolerance when they present a more pronounced curvature. For example, the center line has a more pronounced curvature because of its forefoot and the sheer line, due to its forward part. The presented example is representative of the number of control points and the tolerances that can be achieved with the proposed method. The foremost end of the 3D chine curve will be close to the center line, below the tolerance, but not exactly on the center line because of the fit of the center line points. This is corrected by perpendicularly projecting this point over the B-spline that represents the center line. A spray rail can be obtained in the chine by adding the width of the rail to the Y coordinate of the control points of the chine line (Figure 22, right), with the exception of the control point placed on the center line. This simple approach produces realistic results.

Figure 22. Example of 3D curves and how a spray rail is created.

3.7. Definition of the Stations

The next step is the creation of the stations where the hull surfaces will rest. These curves will be formed from two pieces, one between the center line and the chine and the second one between the chine and the sheer line, because of the topology of the hull. If a spray rail had been created, as in Figure 22 (right), the station would contain a straight segment within the inner and the outer chine lines. Therefore, the first step is to define the ends of the pieces for every station by calculating the intersection points of the 3D curves with different planes Xi, $i = 1, q$, perpendicular to the X-axis and inside the interval $[0, Ls]$.

The intersection points are obtained numerically by subdividing the B-spline curves into Bézier pieces and then proceeding in a similar way as in the previous section to obtain data points for the minimum squares fitting.

The stations of a planing hull can be convex, concave or straight. In addition, they can change their type along the length of the ship. For example, the part of the station under the chine is convex most of the time (Figure 23, left), while the upper part between the chine and the sheer can be a concave curve (Figure 23, right), particularly if the fore part of the ship is designed to expel the water when advancing in bow waves. Thus, the convex/concave aspect of the curve must be controlled. A set of q stations will be defined. For every piece of a station below and above the chine, the method uses a 2nd degree B-spline with three control points:

$$c_d(u) = B_0^2 \cdot P_0 + B_1^2 \cdot P_1 + B_2^2 \cdot P_2; \quad d = 0, q-1$$

The ends of the curves $P_0(YP_0, ZP_0)$ and $P_2(YP_2, ZP_2)$ will be placed in the 3D curves and are calculated as described previously, while the second control point P_1, will be calculated to obtain a given concavity/convexity.

The amount of concavity is set by interpolating a given point, Q. In the proposed method, this point is defined by a given distance to the segment P_0P_2 and by a given position along P_0P_2. The distance is set relative to the length of the segment P_0P_2 and can be positive or negative, i.e. 5% of P_0P_2.

Mathematically, the B-spline $s_i(u)$, interpolates Q at $u = 0.5$. This has the advantage of producing a tangent line at Q that is parallel to P_0P_2 because of the properties of 2nd degree B-spline curves.

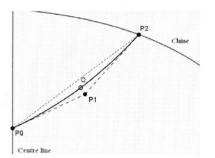 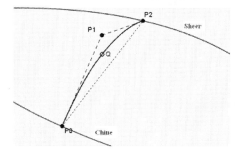

Figure 23. Definition of the stations; convex or concave curves.

In this manner, the deflection of the curve from the segment P_0P_2 is controlled. Moving Q along the length P_0P_2 helps to round the ends of the curve, which is helpful near the center line in the fore part (forefoot), or close to the sheer line in the case of convex stations. The coordinates of P_1 are calculated as follows:

$$P1 = (X_i, 2 \cdot Y_Q - \frac{YP_0 + YP_2}{2}, 2 \cdot Z_Q - \frac{ZP_0 + ZP_2}{2}) \tag{35}$$

Realistic results can be obtained by placing the projection of Q over P_0P_2 at the mid-point of this segment and by using a deflection of -5% ~ +5%, depending on the design characteristics. Examples of the stations that can be obtained with the presented method are depicted in Figure 24. In this example, Q lies in the middle of P_0P_2 for every station and maintains a distance of 3% P_0P_2 both below and above the chine.

The next example (Figure 25) uses the same 3D curves as the prior example, but the position of Q is different for the bow sections. The forefoot of the curves below the chine is a bit more rounded, Q lies at 30% of P_0P_2 at the forward stations and the deflection is maintained at 3%. Over the chine, Q lies at approximately 60% of P_0P_2 with a deflection of approximately -5% for the fore sections.

After the stations have been calculated, the resulting wire model of the ship hull enables hydrostatic and stability calculations to be made with reasonable precision. The distribution of Q along the ship's length is a good indicator of the fairing of the hull. Thus, if the position of Q is plotted for the different stations, the resulting curve should be soft and without bumps or hollows. This will produce faired hull surfaces that will rest on the calculated stations, as will be explained in the next section.

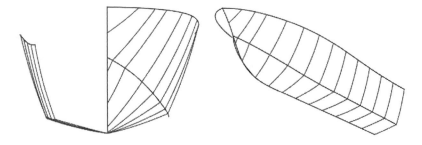

Figure 24. Example of a ship with convex stations over the chine.

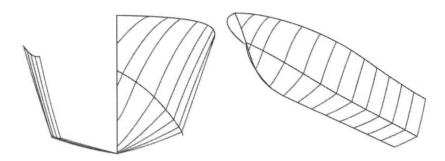

Figure 25. Example of a ship with concave stations over the chine.

4. LOFTING SURFACE OF THE STATIONS

The last step of the method, both for a round bilge or hard-chine hull, is the definition of B-spline surfaces that lean on the stations previously defined, in the case of the hard-chine hull, one surface between the chine and the center lines and a second one between the chine and the sheer lines. If a spray rail exists in this last case, this will produce a third surface between the inner and outer chine lines. The generalization from curves to surfaces is not difficult due to the properties of B-splines, and a lofting surface of the station pieces can easily be defined. Lofting (or Skinning) is one of the most widely used tools for interactive shape description in CAD, with different peculiarities, i.e. [26]. The transition from spline curves to spline surfaces is achieved by turning the control polygon into a control net of control points $W_{ij}(X_{ij}, Y_{ij}, Z_{ij})$, using the same B-spline basis for the two parameters u and v, as well as using two different lists of knots: $\{u_{-1},......,u_{N+n}\}$ and $\{v_{-1},......,v_{M+m}\}$. The lofting process of a set of q B-splines (station pieces) with the same degree and list of knots is as follows: find a B-spline surface S, with degree n by m and $(N+1)$ by $(M+1)$ control points and a list of knots $\{u_{-1},......,u_{N+n}\}$ and $\{v_{-1},......,v_{M+m}\}$ according to Eq. (36) that interpolates q different B-splines c_d $(d=0,...,q-1)$ of nth degree with $N+1$ control points and a list of knots $\{u_{-1},......,u_{N+n}\}$ with the form of Eq. (37).

$$\mathbf{S}(u,v) = \sum_{i=0}^{N} \sum_{j=0}^{M} \mathbf{W_{ij}} \cdot B_i^n(u) \cdot B_j^m(v) \tag{36}$$

$$\mathbf{c_d}(u) = \sum_{i=0}^{N} \mathbf{V_{id}} \cdot B_i^n(u) \quad (d=0,...,q-1) \tag{37}$$

Note that V_{id} are the control points of the different stations obtained in Section 0.4 (Round Bilge) or Section 0.7 (Hard-chine) and expressed in matrix form. Values for N, M and q depend on user preferences regarding the definition of the surface. The interpolation can be written as:

$$\mathbf{S}(u,v_d) = \sum_{i=0}^{N} \left(\sum_{j=0}^{M} \mathbf{W_{ij}} \cdot B_j^m(v_d) \right) \cdot B_i^n(u) = \sum_{i=0}^{N} \mathbf{V_{id}} \cdot B_i^n(u) = \mathbf{c_d}(u) \quad (d=0,...,q-1) \tag{38}$$

This group of equations has to be solved for a set of values of parameter v_d $(d=0,...,q-1)$, termed the choice of the parameterization. The centripetal parameterization produces good results for ship hulls. By identifying equal coefficients for every row of Eq.(38), $i = 0,...N$, the following linear system is obtained:

$$\sum_{j=0}^{M} \mathbf{W_{ij}} \cdot B_j^m(v_d) = \mathbf{V_{id}} \quad (d=0,...,q-1)$$

To obtain a unique solution for this system, $M+1$ must equal q, where q is the number of stations that defines the surface. The $(M+1)$ by $(N+1)$ solutions are the control points W_{ij} of the lofting surface of Eq. (36) containing the stations.

5. APPLICATION EXAMPLES

Two different ship hulls have been generated by following the described method, a round bilge patrol boat, and a hard-chine motor-yacht. The hulls are described by the parameters presented in Table 4 in the case of the round bilge, and in Table 5 in the case of the hard-chine hull, which were selected by the designer according to the design characteristics. The final result is a B-spline surface that rests upon the stations and contains the constraints.

Table 4. Parameters of the application examples

Constraint	Patrol Boat
Disp (m^3)	46.6
LCB (m)	7.41
XMAX (m)	7.65
α_I, α_F (°)	17°, 32°
LWL (m)	16.7
AWP (m^2)	81.8
LCF (m)	7.29
A'$_I$, α'_F (°)	14°, 60°
T (m)	0.89

Table 5. Parameters of the examples

Name	Motor-yatch
Ls	124.0
L0	66.5
Lx	62.8
Lc	114.4
Xc	49.3
X_{C1}	55.4
Bs	11.1
Bx	13.7
Bc	10.2
Sp	1.2
Hs	16.3
Hc	11.2
hr	0.0
hs	12.8
hc	3.3
Z_{C1}	4.6
α_K	29
α_S	129
β'_S	2
α'_S	1
α_C	23
β_C	1
α'_C	8
β'_C	0
$2 \cdot Ac$	2049

Example 1: Round Bilge Patrol Boat

This example shows the lines of a fast patrol boat. The SAC and waterline can be seen in the top of Figure 26. Based on these curves, the stations of the hull are presented below. This example shows a ship hull that has straight sections in the lower part of the hull to improve the dynamic lift of the ship. The design has concave sections at the above-water fore part of the ship to expel the spray drag created when the ship advances at fast speed.

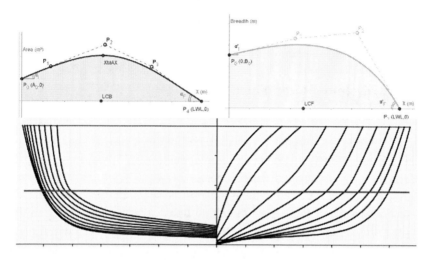

Figure 26. Round Bilge Patrol Boat.

Figure 27. Rendered view of the patrol boat hull.

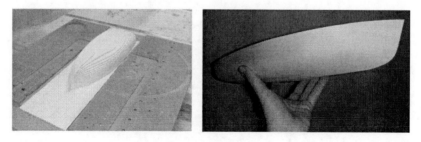

Figure 28. Constructed model of the fast patrol boat.

The stations present a distribution of dead-rise angles β along the ship's length. The tolerances (2 mm) of the cubic B-spline curves that model the stations, as detailed in section 0, were achieved with 9 control points. The bounding box for this example is [16820 mm, 6690 mm, 2000 mm]. According to section 3.5, constructing the lofting surface of such stations defines the constrained ship hull (Figure 27). In this figure, the underwater portion of the hull that holds the displacement and LCB is shaded in grey, and the above-water part of the ship that joins the waterline with the deck line is shaded in a dark colour. The surface information was transferred to a numerical control machine and a scaled prototype was constructed in wood. This was done to physically test the quality of the surface and its ability to serve as a guide for construction.

Example 2: Hard-chine Motor Yatch

This example shows a 126-ft (37.8 m) fast motor-yacht. Notice the convex forward sections (Figure 29) and the absence of warp at the bottom. Dead-rise is 18° in the transom, and the keel is flat for an undisturbed center-jet inlet and for lift. According to Section 0.7, the maximum deflection of the stations varies linearly between 1% ~ 2% for the aft and fore sections, and the deflection is placed in the middle of the segment P_0P_2 (u=0.5).

The projections of the center, sheer and chine lines calculated according to the method, are shown at the top of Figure 29, while a 10-station front view is depicted in the middle of the figures, according to Section 0.7. The 3D lines were calculated by considering 80 points for every 3D curve and adjusting those points with a B-spline curve to obtain a maximum tolerance of 0.01-feet, following Section 0.4. The lower part of the figures shows rendered views of the lofting surfaces of Section 0.

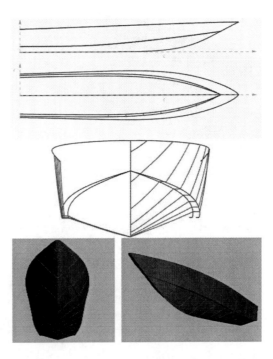

Figure 29. Hard-chine motor yacht.

CONCLUSION

This chapter has presented a practical method for defining a constrained surface for a ship hull according to normal design parameters for a naval architect. This constrained definition enables rapid changes in the design procedure and avoids unnecessary manual work because a valid hull form can be automatically generated and geometrically constrained to the design parameters.

The facility to change the characteristics of the ship hull by modifying numerical parameters rather than by manipulating tens of surface control points is beneficial and was the motivation behind this research. However, constrained definition techniques are quite complex and they increase in difficulty as more detailed features are included in the design. Every constrained definition technique is only capable of representing a limited range of hull forms compared with the total flexibility inherent in a manually defined hull. The method presented here is not an exception, but it is a good compromise when creating simple hull forms that can be manually modified at a later time to add complex features.

The method uses different parameters in the case of a round bilge hull, and in the case of a hard-chine hull. In the first case, the method starts by creating a SAC and a waterline that contains most of the constraints. Then, a B-spline is produced that satisfies the desired constraints by solving a nonlinear set of equations for both curves. Other methods use optimisation techniques that may require additional parameters and normally do not use a waterline in the definition. The presented method constrains the angles at the ends of the mentioned curves, providing the designer with control over the volume/area distribution along the ship's length.

This method produces offset curves (stations) that fit the parameters of the previous curves. This is done analytically for the underwater part of the ship, resulting in more realistic ship hulls. A 2nd degree Bézier curve is used for the above-water part of the ship. This allows the waterline to be softly joined to the deckline and allows concave/convex sections to be obtained. The use of analytic expressions to define the ship hull is uncommon. Most other methods require a library of hull stations or other lines of the hull.

Points on the analytic and Bézier curves are merged with a B-spline using least square fitting, considering a distance tolerance into its parameterisation. This has an important influence from the construction point of view. The number of longitudinal and transverse control points on the surface can be defined, so the definition of the surface can be increased and then manually modified in later stages to include appendages or other design requirements, which may be difficult to include in the initial definition of a constraint design procedure.

In the case of a hard-chine hull, the method is based on projections of the significant lines of the hull including the center, sheer and chine lines. These lines were defined based on geometrical constraints without the use of templates. The transformation of a template in a planing hull can change important parameters of this type of ship, such as the dead-rise angle. This can alter the performance compared to the original template.

The method presented here can reproduce specific features of hard-chine hulls, such as a straight sheer or a center line with some rocker. The method can also include the enclosed area and its centroid into the chine definition, which are important parameters related to the operation of the ship.

After these 2D lines have been defined, corresponding 3D curve were defined using a least squared fitting that considered a distance tolerance in its parameterization. The stations of the design can include a certain amount of convexity/concavity that is different based on the type of design (e.g., fishing vessels, recreational crafts).

The final result is a B-spline hull that can be easily exported to specialised computer programs for further stages of the design process. The same hull representation can be used for the design and in all subsequent stages of the process of developing the ship.

REFERENCES

[1] *Proceedings of 31st WEGEMT School*. (1999). CFD for ship and offshore design. Hamburg.

[2] *Proceedings of Marine CFD 2008*. Rina. London.

[3] Kuiper, G. (1970). Preliminary Design of Ship Lines by Mathematical Methods. Journal of Ship Research, 14, 52-66.

[4] Reed A. M., Nowacki H. (1974). Interactive Creation of Fair Ship Lines. *Journal of Ship Research*, 18, 96-112.

[5] Creutz, G., Schubert, C., (1978). *Interactive Curve Creation from Form Parameters by Means of B-splines*, Schiffstechnik 25.

[6] Keane, A.J., (1988). A computer based method for hull form concept design: applications to stability analyses. *Transactions RINA*, 130, 61–75.

[7] Yilmatz, H., Kukner, A. (1999). Evaluation of cross curves of fishing vessels at the preliminary design stage. *Ocean Engineering*, 26, 979–990.

[8] Mancuso, A. (2006). Parametric design of sailing hull shapes. *Ocean Engineering* 33, 2006, 234-246.

[9] Kim, H.C. (2004). *Parametric design of ship hull forms with a complex multiple domain surface topology*. Ph.D. thesis. Tech. Universität Berlin, Berlin, Germany.

[10] Bole, M. (2002). *A hull surface generation technique based on form topology and geometric constraint approach*. Ph. D thesis. University of Strathclyde, UK.

[11] Nowacki, H. (2010). *Five decades of Computer-Aided Ship Design*, CAD 42, 956-969.

[12] Calkins, D. E., Schachter, R. D., Oliveira, L. T. (2001). An automated computational method for planing hull form definition in concept design. *Ocean Engineering* 28, 297-327.

[13] Savitsky, D. (1964). Hydrodynamic Design of Planing Hulls, *Marine Technology*, 1.

[14] Savitsky, D., Brown, W. (1976). Procedures for Hydrodynamic Evaluation of Planing Hulls in Smooth and Rough Water. *Marine Technology* 13.

[15] Blount, D.L., Codega, L.T. (1992). Dynamic Stability of Planing Boats. *Marine Technology*, 29, 4-12.

[16] Clement, E., Blount, D., (1963). Resistance Tests of a Systematic Series of Planing Hulls, *Transactions of the SNAME*, 71.

[17] Lamb, T., (2003). Ship Design And Construction, Chapter 9. Ed. *Society of Naval Architects and Marine Engineers (SNAME)*.

[18] Lewis, E. V., (1987). Principles of Naval Architecture, 1, Ed. *Society of Naval Architects and Marine Engineers (SNAME)*.

[19] Farin, G., (2001). *Curves and Surfaces for CAGD*, Ed. Morgan Kaufmann, San Francisco.

[20] More, J., Burton, G., and Hillstrom, K., (1980). User guide for MINPACK-1, Argonne National Labs Report ANL-80-74, Argonne, Illinois.

[21] Jorde, J-H., (1997). *Mathematics of a Body Plan*, The Naval Architect, 38-42.

[22] Piegl, L. A.; Tiller, W., (1997). *The NURBS Book*, Ed. Springer, 410-413.

[23] Hoschek J., (1988). Intrinsic parametrization for approximation, *Computer Aided Geometric Design* 5, 27-31.

[24] Jenkins, A.; Traub, J., (1970). A three-stage variable-shift iteration for polynomial zeros and its relation to generalized Rayleigh iteration, *Numerische Mathematik* 14, 252-263.

[25] Saunders H. E., (1957). *Hydrodynamics in Ship Design*, sec. 77. Ed. SNAME.

[26] Woodward, C.D., (1988). Skinning techniques for interactive B Spline surface interpolation. *Computer-Aided Design,* 20, 441-451.

In: Ships and Shipbuilding
Editor: José A. Orosa

ISBN: 978-1-62618-787-0
© 2013 Nova Science Publishers, Inc.

Chapter 2

DESCRIPTION OF A PROPULSION SYSTEM OF A SHIP DESIGNED TO TRANSPORT LNG WITH A STEAM PLANT

A. De Miguel Catoira[1], J. Romero Gómez, M. Romero Gómez and R. Bouzón Otero
Department of Energy and Marine Propulsion
University of A Coruña, Spain

ABSTRACT

The chapter presents a description of the propulsion system of a ship designed to carry liquefied natural gas (LNG). This system is installed in a high percentage of these vessels within the world fleet. The installation described comprises a steam plant with two boilers and a high and low pressure turbine.

The necessary auxiliary systems for the correct operation of the installation are depicted. Such auxiliary systems include the condensate system, whose purpose is to condense the installation's steam as well as increase the water temperature in order to improve the thermal efficiency of the plant.

Another is the feed water system, whose function is to increase the pressure of the water to be introduced into the boiler. Moreover, results are shown of the energy balances and fuel consumption for different vessel running conditions and an analysis of the operating condition at 90% MCR using heavy fuel oil, natural gas or both concurrently, as fuel is carried out.

Also investigated is the plant during the processes of loading and discharging. The conclusion derived was that the thermal efficiency of the plant is low when compared with other propulsion systems already installed, which generates a high cost of operating the vessel. Notwithstanding, the system stands out in its simplicity, its ease of operation and low maintenance, resulting in such expense to be very low compared to other propulsion systems.

---

[1] Email: jaorosa@udc.es.

INTRODUCTION

One of the parameters of greatest influence when determining the viability of propulsion plant type on a merchant vessel is the operating cost, with fuel cost being one of the most important. So influential is this factor that, depending on the type of ship, an increase in fuel prices can result in it no longer being competitive within the market. It is therefore essential to head investigations towards reducing these costs when defining a project. The price of marine fuels is directly and proportionately linked to the price of crude oil [1]. Until 1973, fuel cost was minimal and therefore vessels did not consider energy saving. Engine power and simplicity of operation and maintenance was sought rather than energy efficiency. Few stage team turbines were used, boilers without economizers, short strokes diesel engines that burnt light fuel, simple turbochargers, engines without superchargers, steam deck machinery, gas turbines, turbo generators and a wide range of low energy efficiency equipment. The increase in oil prices led to the rapid abandon of these equipments on merchant ships. In the late 70s there was a fixation on fuel economy within the world's largest shipping companies. With huge investments propulsion turbines were changed, replacing them with diesel engines in large container ships as well as on cruise ships such as the Queen Elizabeth II and some oil tankers. Many vessels with steam plants were recycled. Thereafter, merchant vessels opted for diesel engines rather than steam. Since the 90s, new constructions installed with a steam plant were LNG vessels, an installation similar to that described in [2-4] and in this chapter. In late 2000s, a steam system was proposed for ships carrying coal [5]. This system would use a complex steam cycle, using the coal being transported by the ship as fuel, which is burnt in the boilers. The technical and economic feasibility of this system is studied, concluding that the project is viable with current oil prices, with it being more cost effective than the consumption of heavy fuel oil. The environmental effects of coal consumption, however, should be taken into account, as international regulations are becoming increasingly restrictive regarding the consumption of coal.

1. CURRENT LNG STATUS

In recent years the maritime trade that has undergone most growth has been LNG [6]. This has resulted in research to improve the efficiency of propulsion systems being focused on these types of vessels. In fact, many manufacturers have centred their developments in this field [7-9]. [10] analyses of the different propulsion systems on LNG vessels, such as:

- Steam turbine with mechanical propulsion.
- Dual diesel engines with electric propulsion.
- Gas turbine with electric propulsion.
- Dual diesel engines with mechanical propulsion.
- Diesel propulsion with re-liquefaction plant.

The chapter evaluates aspects related to maintenance and safety, with preference to the availability of the system as the most important feature, over costs, initial investment, operation, maintenance or others, due to the special characteristics of this type of vessel. The comparison between LNG ship propulsion systems is discussed in [11] and [12]. These comparative papers illustrate the comparison from an economic standpoint, in order to evaluate the costs of installation and operation of the chosen propulsion system. In [11] an algorithm is developed to compare these costs between two different systems. This allows you to see which is more cost effective, being able to vary the requirements to which the ship is subjected as well as the price of fuel, so that a range of different situations can be investigated. The above references are related to the propulsion systems of ships. An important aspect, however, which many authors and researchers are working on, is the complete or partial reliquefaction of boil off gas (BOG) generated in the vessel. [13] and [14] carry out a descriptive study and analyse the control system of reliquefaction systems installed on LNG ships . These systems are to have the capacity to reliquefy all BOG produced, since the propulsion system consists of conventional diesel engines burning heavy fuel oil. Due to the high specific consumption of these plants, the operation and design of these should be adequate to ensure reasonable energy consumption, and achieve to re-liquefaction of all the BOG produced. [15] carries out an exergy analysis of the plants analysed in [13] and [14], obtaining all the fundamental parameters upon which the exergetic efficiency depend.

2. DESCRIPTION

This section describes each of the systems that the plant comprises. This may be divided into sections that are shown in the following figure:

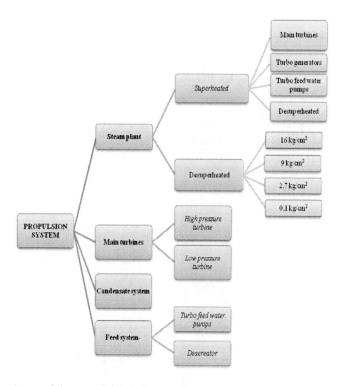

Figure 1. General scheme of the propulsion system.

3. STEAM PLANT

3.1. Superheated Steam Plant

To generate superheated steam, steam at high pressure and high temperature, this type of ship is provided with two boilers with the following technical specifications:

- Features: superheater, a high steam collector and a low water one.
- Maximum flow of superheated steam: 65,000 kg/h.
- Regular flow of superheated steam: 50,000 kg/h.
- Properties of superheated steam: 62 kg/cm^2 at 515 °C.

These features may vary depending on the manufacturer and type of boilers installed. All ship's steam requirements are covered by these two boilers. Regulation of superheated steam temperature is achieved with a control valve, which connects the water drum and the superheater. This control valve allows the passage of wet steam from the water drum to the superheater to temper the superheated steam to the assigned temperature.

Part of the superheated steam is diverted to the desuperheater located in the steam drum to achieve a temperature of between 280 and 320 °C. The main steam valves feeding the turbine also connect both boilers. A similar arrangement is seen in the valves that feed the turbo-generators and turbo-pumps. These main valves have bypass valves for heating and line drain valves. The systems to which superheated steam is supplied are as follows:

- Main Turbine.
- Turbo generators.
- Turbo feed water pumps
- Desuperheated steam system.

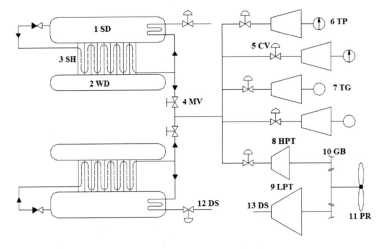

1 Steam Drum. 2 Water Drum. 3 Super-Heater. 4 Manual Valve. 5 Control Valve. 6 Turbo pump. 7 Turbo Generator. 8 High Pressure Turbine. 10 Gear Box. 11 Propeller. 12 to Desuperheated Steam system. 13 from Desuperheated Steam system

Figure 2. Super-heated steam system 60 kg/cm^2.

3.1.1. Desuperheated Steam System

To obtain desuperheated steam at a pressure of 60 kg/cm^2 and a temperature ranging between 280-320 °C, depending on the load of the boiler, desuperheated steam is diverted after the safety valve at the outlet of the superheater, toward the steam manifold through a valve for both boilers.

Once desuperheated in the manifold, it is distributed to various services through the automatic valves. The desuperheated steam consuming services are as follows:

- The main low pressure turbine emergency steaming connection.
- Dump steam system.
- The 16 kg/cm^2 steam system through 60/16 kg/cm^2 reducing valve.
- The 9 kg/cm^2 steam system through 60/9 kg/cm^2 reducing valve.
- Boiler soot blowers.

Another type of superheated steam which the ship features is that of 9 kg/cm^2. This line has two external desuperheaters, which are used to lower the steam temperature to 190 °C using feed water injected as a spray for tempering.

The vessel's control system controls the pressure of the system's desuperheated steam of 16 and 9 kg/cm^2. The temperature of the auxiliary steam systems is controlled by means of signals from the temperature transmitters installed in the lines. The general disposition of the desuperheated steam system is shown in figure 3 below.

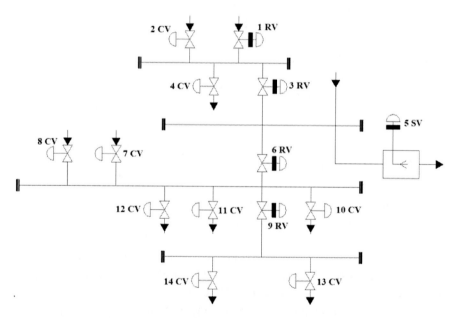

1 Reduction Valve 60/16 kg/cm^2. 2 Control Valve from high pressure bleed. 3 Reduction Valve 16/9 kg/cm^2. 4 Control Valve. 5 Saturation Valve to 9 kg/cm^2 low temperature. 6 Reduction Valve 9/2.7 kg/cm^2. 7 Control Valve from medium pressure bleed. 8 Control Valve. 9 Reduction Valve 2.7/0.1 kg/cm^2. 10 Control Valve. 11 Control Valve. 12 Control Valve. 13 Control Valve. 14 Control Valve.

Figure 3. Desuperheated steam system, 16, 9, 2.7 y 0.1 kg/cm^2.

16 kg/cm^2 Steam System

The 16 kg/cm2 steam system is fed by the high pressure bleeding steam from the main turbine, or by 60/16 kg/cm^2 reducing valves. As the speed of the turbine increases, so does pressure in the high pressure bleed, which is why the upper bleeding valve opens, feeding the line with 16kg/cm2. The 16 kg/cm^2 line supplies steam to the following services:

- Main condenser and evaporators´ air ejector.
- 16/9 kg/cm^2 supply valve.

The 16/9 kg/cm^2 reduction valve only operates in situations in which the high pressure bleed puts too much steam into the 16 kg/cm^2 system and the 60/16 kg/cm reduction valve is closed.

9 kg/cm^2 High and Low Temperature Steam System

The 9 kg/cm^2 steam system can be fed both with the 60/9 kg/cm^2 reduction valve, as with the 16/9 kg/cm^2 reduction valve, which are controlled by the vessel's control system.

9 kg/cm^2 High Temperature Steam System

The 9 kg/cm^2 high-temperature steam system supplies the following services:

- Drain steam for tubes of fuel oil boiler burners.
- Atomisation steam for the boiler burners.
- Steam for fire prevention of the boiler air register.
- Steam for the 9/2.7 kg/cm^2 reduction valve.
- 9 kg/cm^2 low temperature steam.

This line is fed by the 16 kg/cm^2 system or the 60 kg/cm^2 desuperheated system through reducing valve both. The operator must ensure that the 60/9 kg/cm^2 valve begins to close as the 16/9 kg/cm^2 begins to open and vice versa. In order to avoid interference and oscillations between the valves of 16/9 kg/cm^2 and the 60/9 kg/cm^2, the following pressures are established in the control system:

- From the 16 kg/cm^2 system: 9.3 kg/cm^2.
- From the 60 kg/cm^2 system: 8.8 kg/cm^2.

Given that this 9 kg/cm^2 line feeds the boiler spray system, as well as other systems of the boiler, is not convenient to stop the boiler, unless a fault or incident occurs.

The atomisation system's lines have multiple connections to the 2.7 kg/cm^2 system in order to eliminate the condensation of steam in the line when the system is stopped for a long period of time, which would result in the entry of water to the burner and could cause flame failure.

3.1.2. Low Temperature 9 kg/cm^2 Steam System

This system is powered from the high temperature 9 kg/cm^2 steam line through the external desuperheaters. The temperature is reduced to 190 °C with valves that inject water from the feed water system, which has too high a pressure. For this reason it is reduced to 20 kg/cm² through a control valve. This type of valve prevents its complete closure, avoiding a sharp rise in the pressure when it returns to open.

As can be seen in figure 4, the system is used for the following services:

- Heating services to tanks and other services
- Atomising steam for the inert gas generator and heating steam.
- Incinerator services.
- Cleaning cylinder of the burner rods.
- Oil Separator.
- Air conditioning system.
- Steam-cleaning system of seawater suction filters.
- Tracing steam.
- Deck services.
- Heating for atmospheric observation tank.

The system is divided into several individual sections, each with their respective isolation valves in order to perform maintenance operations if necessary. The following two sections have pressure reducing valves:

- Sewage treatment plant heaters.
- Oil separator and air conditioning.

Many of the system's services, such as the heating of heavy fuel oil tanks, have an autonomous system for temperature control through the use of thermostatic valves. It is for this reason that they are not controlled by the ship's control system.

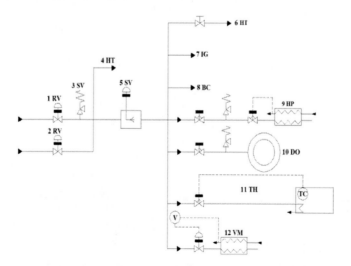

1 Reduction Valve 60/9 kg/cm². 2 Reduction Valve 16/9 kg/cm². 3 Safety Valve. 4 To 9 kg/cm² High Temperature. 5 Saturation Valve. 6 Heating Tanks. 7 Inert Gas atomising. 8 Burner Cleaning. 9 Heating Purifier. 10 Deoiler. 11 Tanks Heating. 12 Viscosimeter.

Figure 4. 9 kg/cm² low temperature steam system.

3.1.3. Medium Pressure Exhaust System

The medium pressure exhaust system, as shown in figure 5, operates at 2.7 kg/cm^2 and is fed from three separate sources:

- Medium pressure bleed of the propulsion turbine.
- Turbo pumps steam discharge.
- Reducing valve 9/2.7 kg/cm^2.

The steam supplied to this system comes mainly from the turbo-pump discharge. While the plant operates at steady state, one of the two turbo pumps is operational. The discharge of these turbines is attached in a line which feeds the medium pressure steam system.

The medium pressure bleed coming from the junction of the high and low turbine operates similarly to high bleed, as described above. If the turbine speed increases, so will the pressure of the line joining the high and low turbine. When the latter reaches 3.8 kg/cm^2, the valve opens, and would close in the event that the pressure descended to 3.3 kg/cm^2.

A pressure control valve is installed after the bleed valve in case the pressure of the joining line of the high and low turbine exceeds the system's nominal working pressure. The bleed is continuously purged to the main condenser and for this the line has a calibrated orifice located before the automatic valve.

It could be the case that there is not enough steam to supply the system, so that steam would be delivered through the pressure reducing valve, which has a set point of 2.7 kg/cm^2. This valve is regulated by the control system through the signal sent by the pressure sensors. The opposite case could arise: that the steam supply to the system exceeds the demand. This excess of pressure is relieved to a condenser through a valve with a set-point at 2.9 kg/cm^2.

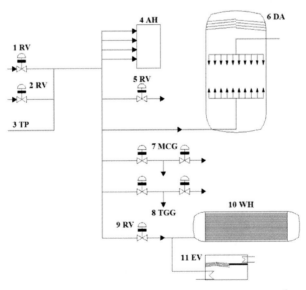

1 Reduction Valve from medium pressure bleed 2 Reduction valve from 9 kg/cm^2 3 Turbo Pump Discharging 4 Air Heater 5 Reduction Valve to main condenser 6 Dearator 7 Main Condenser Glands 8 Turbo Generator Glands 9 Reduction Valve 0.1 kg/cm^2 10 low pressure feed Water Heater 11 Evaporator.

Figure 5. Medium and low pressure steam exhaust systems.

If the main condenser were out of service, the auxiliary condenser would have to be used. This procedure can be performed automatically by the pressure relief control valve, as it can allow steam to pass towards the main or auxiliary condenser, by operating the respective valves. The medium pressure exhaust steam system feeds the following services:

- Deaereator heating.
- Main turbine heating steam.
- Steam for the glands of the main turbine and turbo-generators.
- Boiler air heaters.

The purpose of heating steam to the deaereator is to increase the temperature of the condensate and the release of uncondensables introduced into the system. The heating of steam to the main turbine is to achieve a slow and continuous heating of the turbines and their housings. This is performed by introducing medium pressure exhaust steam into the inlet of the high pressure turbine and it continues through the low turbine to the condenser.

Each line of gland steam for the main turbine and the turbo generators has a calibrated orifice that empties directly to one of the condensers. This prevents the condensation of water in case the system is not used during a long period of time and there is steam flow when the equipment is running. Air heaters are heat exchangers of finned tubes that require little maintenance and simply require checking that the purges are (open) and that the air acquires the correct temperature. The differential pressure between the inlet and outlet should also be monitored to check if they are dirty.

3.1.4. Low Pressure Exhaust System

The low pressure evacuation system operates at a pressure of 0.1 kg/cm^2 and is fed from two separate systems:

- Low-pressure bleed.
- 2.7/0.1 kg/cm^2 regulating valve.

The low pressure bleed comes from the third stage of the lower turbine and is controlled with an automatic valve which opens when the medium bleed reaches 3.8 in pressure and closes when it drops to 3.3 kg/cm^2. It is important that the bleed valve is closed when the load is low and opens when the load increases. If this non-return valve were to work incorrectly, the main condenser vacuum could be lost. If the steam supplying the lower bleed were insufficient, or if it were is closed, the system can supply steam through the 2.7/0.1 kg/cm^2 reduction valve. This valve is commanded from the control system according to the signal received from the pressure sensors installed. The low pressure steam exhaust system feeds the following services:

- Heating for the evaporators.
- Low pressure feed water heater.

Both evaporators have a control valve supplying steam to these, as well as some attemperators to reduce its temperature by injecting condensate.

The low pressure feed water heater is used to heat the condensate water before it enters the deaereator. The purpose of the heater is to improve the thermal efficiency of the plant.

4. MAIN TURBINE

The main propulsion turbine consists of a ten-stage high turbine, an eight-stage low turbine, combined with an astern turbine, a maneuvering valve, a main condenser and a gear box reducer.

- High pressure turbine: one Curtis stage (for regulating) and 8 Rateau.
- Low pressure turbine: 4 Rateau stages and 4 reactions.
- Astern turbine: 2 Curtis stages.
- Power: Maximum 28000 kW: (39 nozzles): normal 25 200 kW: (31 nozzles) (these data are approximate, according to the model and manufacturer).

The high pressure turbine and the maneuvering valve are mounted on a common frame. The low is mounted directly onto the main condenser. The reductive is of double helical type and articulated double reduction. It is installed in the stern of the main turbine. The main thrust bearing is installed on the bow side of the reducer.

Turbine power is transmitted to the propeller shaft via a flexible coupling reducer. The total power is controlled by the throttling action of the maneuvering valve which is controlled by a hydraulic servo mechanism.

During forward navigation, the high pressure steam, through the maneuver valve, enters the high pressure turbine through the front part and flows towards the stern. Its exhaust is aimed at the low-pressure turbine, with the steam flowing from bow to stern. The lower turbine exhaust enters directly into the main condenser.

In order to reverse, the high pressure steam coming from the maneuvering valve enters the astern turbine, located at the end of the low turbine discharge, and flows toward stern stator part. The exhaust steam enters directly into the main condenser.

In case of an emergency operation, connections are fitted with emergency steam. In order to work with only the high pressure turbine, an emergency evacuation connection is prepared from the crossover (connection between high and low turbine) to the main condenser and thus the exhaust steam of the high pressure turbine can be evacuated directly to the condenser.

In order to operate only with the low-pressure turbine, an emergency steam connection in the transversal line is provided, thus supplying steam directly from the boiler.

5. CONDENSATE SYSTEM

From here on, "condensate" will refer to all steam that has completed its process and has changed in state to its liquid phase in any of the condensers (main, atmospheric or in that of the gland).

The basic function of the condensate system is to extract liquid from the condensers and take it to the deaereator, so that it can then be directed to the boilers through the feed pumps as feed water. The condensate system has a network joining with tanks of distilled water, which are the circuit water reserves. The whole system is shown in figure 6.

For the extraction of condensate, there are two main or high-capacity pumps and a low-capacity pump. These pumps are self-cavitating: they lose their suction if there is not the adequate amount of pressure in the suction. Cavitation causes serious damage and erosion in centrifugal pumps, primarily in its impeller.

It should also be noted that its suction pressure is not positive. To avoid this cavitation problem, these pumps are designed to prevent damage as a result of cavitation in the impeller. The condensate removal pumps drive the water from the main condenser to the gland steam condenser coils and to the low pressure feed water heater.

The course of the condensate from the main condenser has two essential purposes: condensing or cooling other residual heat sources and raising the temperature of the condensate before entering the boilers. The purpose of both is to increase the thermal efficiency of the plant. There is a deviation in the condensate line after the steam condenser output which goes to spray nozzles where the steam relief lines are discharged. This reduces the excess pressure of the steam system and causes tempering of the steam in the main condenser itself.

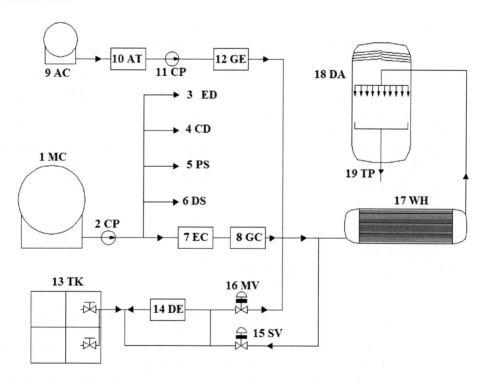

1 Main Condenser 2 Condensate Pump 3 Evaporator Desuperheater 4 Chemical Dosing 5 feed water Pump Sealing 6 Desuperheater 7 Evaporators Condensers 8 Gland Condenser 9 Atmosferic Condenser 10 Atmosferic Tank 11 Condensate Pump 12 Grease Extractor 13 Tanks 14 Deionizer 15 Spill Valve 16 Make-up Valve 17 low pressure feed Water Heater 18 Dearator.

Figure 6. Condensate system.

There are other lines that deviate from the condensate system to feed the following services:

- Spray to temper steam to the evaporators.
- Filling of the chemical dosing tanks for boiler water.
- Unit obtaining water samples for analysis.
- Spray for external tempers of the steam discharge line.
- Water for mechanical seals of the feed water pumps.

Once the condensate has been through all these elements, it proceeds to the circuit deaereator, which performs the following functions:

- Store all condensate returning from the boilers.
- Act as a feed water heater.
- Eliminate incondensable gases dissolved in the feed water.
- Feed water reserve for boilers in case of the fall of the plant.

The deaereator is located at the highest point of the whole plant to generate a positive suction pressure in the feed pumps and larger than the saturation pressure at the inlet of the latter. This avoids cavitation in the supply pump.

Water enters the deaereator through a series of spray nozzles, where the heating steam also enters. Introducing the condensate as a spray ensures good contact with the steam, which enters at a pressure of 2.7 kg/cm^2. It behaves as a mixing heat exchanger: the steam condenses to join the condensate and to become feed water.

During this process, oxygen is released along with other uncondensable gases, which are vented through the top of the deaereator via a valve that should always be properly calibrated to ensure optimum performance of the deaereator. The deaereator water level is maintained by the level transmitter, which controls two valves: the spill valve and make-up valve, which function according to a range of levels.

6. FEED WATER SYSTEM

The condensate from the plant´s steam consumers returns to the main feed water system from the plant to the boilers, as shown in figure 7. This implies that an increase in pressure is required in order to create a given flow to the steam collectors of the boiler. The feed water turbo pumps are responsible for this. These are of centrifugal, high speed and multistage in type. The system has two of these pumps installed, one of which supplies water to the boilers, while the other remains on standby.

The system is also fitted with a third electrically-driven pump, for the start and stop of plant. There is another cold start electric pump with the same purpose as the latter, except that it can only be moved by a motor powered by the emergency switchboard. To move the largest electric auxiliary pump a turbo generator or diesel generator needs to be running.

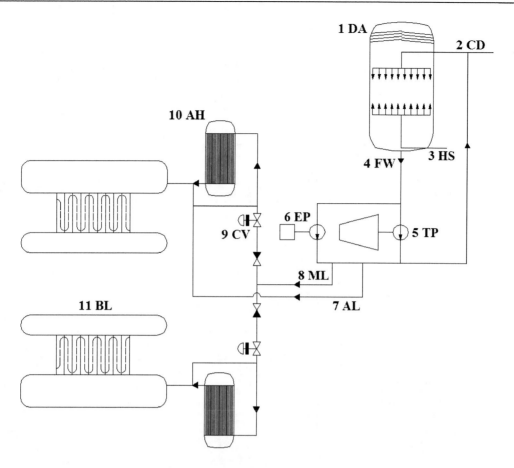

1 Dearator 2 Condensate 3 Heating Steam 4 Feed Water 5 Turbo Pump 6 Electrical Pump 7 Auxiliary Line 8 Main line 9 Control Valve 10 Air Heater 11 Boiler.

Figure 7. Feed water system.

All of these feed pumps have a common suction, directly at the water outlet of the deaereator. The deaereator is located at the highest part of the plant to facilitate removal of the incondensable gases and to provide a sufficient water height for the pumps not to suffer the effects of cavitation. Moreover, in case of a fall of the plant, the deaereator becomes an important water reservoir to keep feeding the boilers.

All pumps have a bifurcation at their disposal that goes to the main or auxiliary feed line. The main feed line is the normal path taken by the water towards the boiler's steam collector. From the pump discharge, it goes towards the economiser inlet head, passing firstly through the supply control valve and the main power motorised valve. The feed water in the economiser increases its temperature due to heat transfer conducted by the boiler exhaust gases. The main feed valve of the boiler is located between the economiser and the steam drum.

The auxiliary feed line's role is to continue to provide water to the boiler, in case of problems arising in the main line or in the feed regulation valve.

The auxiliary feed line valve is operated manually. In normal conditions and under a normal load, it needs a small aperture for maintaining the water level in the steam drum.

Table 1. Performance parameters in different conditions

Parameter/ condition		90% MCR Dual	90% MCR Gas	90% MCR Fuel	Dumping	Loading
Turbine	Shaft output (kW)	25.200	25.200	25.200	-	-
	Shaft revolutions (RPM)	80.1	80.1	80.1	-	-
	Steam pressure (bar)	61	61	61	-	-
	Steam T^{a2} (°C)	510	510	510	-	-
	CondenserVacuum (mmHg)	727	727	727	-	-
Boiler	Steam pressure (bar)	62.8	62.8	62.8	64.3	62.8
	Steam T^a (°C)	515	515	515	515	450
	Feed water T^a (°C)	138	138	138	138	138
	Boiler efficiency	85.8	84.2	88.7	83.9	84.2
Fuels	Heavy fuel oil LHV[3] (kJ/kg)	42.970	-	42.970	-	42.970
	HFO consumption (kg/s)	2.836	-	7.304	-	1.306
	HFO cons. Rate (g/kWh)	112.4	-	289.8	-	-
	Fuel gas LHV (kJ/kg)	52.295	54.160	-	53.545	-
	FG consumption (kg/s)	3.665	6.070	-	3.665	-
	FG cons[4]. Rate (g/kWh)	145.5	240	-	-	-
	Global efficiency (%)	30.75	29.46	30.37	-	-
Others	Electric load (kW)	1.580	1.700	1.280	1.100	2.550
	Sea water T^a (°C)	24	24	24	27	24
	E.R.[5] air T^a (°C)	45	45	45	45	45
	Boil off rate (%/day)	0.15	0.15	0.15	0.15	0.15
	Distilled water production (Tn/day)	30	28	34.1	-	-

7. RESULTS

Table 1 depicts the different operating parameters of the plant for the various conditions of the ship. As can be observed, the vessel's consumption is high given the power output. In effect, specific fuel consumption is high in absolute terms and, conversely, the thermal efficiency of the plant is low (around 30 %). This efficiency is much lower to that of similar vessels fitted with dual diesel engines with an electric drive. The other parameters shown are within normal operating margins in the case of steam turbine installations for the propulsion of merchant ships.

CONCLUSION

Following the description of the system and the results shown, from a thermodynamic stand point, the plant has a low efficiency and is not suitable to be installed due to high operating costs and the environmental impact resulting from the high emissions produced.

---

[2] T^a – Temperature.
[3] LHV – Low heat value.
[4] Cons. – Consumption.
[5] E.R. – Engine room.

On the other hand, the equipment is simpler than that of vessels fitted with dual motors and electric propulsion. In this type of equipment, maintenance costs are higher because there is redundancy in the equipment of such vessels: their technology is more modern and is being developed.

Due to high operating costs, fuel costs and low efficiency, construction of these vessels has diminished. Therefore, these facilities require the development of a more efficient technology, making propulsion with steam turbines profitable once more.

REFERENCES

[1]   Economía y globalización. http://blogs.ua.es/miqueltari/2010/04/24/ evolucion-precio-petroleo. (Accessed December 2012).

[2]   Haglind, F. (2008). A review on the use of gas and steam turbine combined cycles as primemovers for large ships. Part I: Background and design. *Energy Conversion and Management* 49, 3458–3467, 2008.

[3]   Haglind, F. (2008). A review on the use of gas and steam turbine combined cycles as prime movers for large ships. Part II: Previous work and implications. *Energy Conversion and Management*, 49, 3468–3475.

[4]   Haglind F. (2008). A review on the use of gas and steam turbine combined cycles as prime movers for large ships. Part III: Fuels and emissions. *Energy Conversion and Management* 49, 3476–3482.

[5]   Rodriguez, C., Antelo, F., De Miguel, A., Carbia, J. (2011). Study of possibilities of ussing a steam plant type "reheat" and mixed boilers of coal and fuel-oil for the propulsion of bulkcarriers. *Journal of Maritime Research,* 8, 2, 3-12.

[6]   SIGTTO. http://www.sigtto.org/Publications/Publications-and-downloads. (Accessed December 2012).

[7]   MAN BandW Diesel. (2012). LNG Carrier Propulsion by ME-GI Engines and/or Reliquefaction. MAN BandW Diesel A/S, Copenhagen, Denmark.

[8]   MAN BandW Diesel. (2012). Propulsion Trends in LNG Carriers Two-stroke Engines. MAN BandW Diesel A/S, Copenhagen, Denmark.

[9]   Hansen, J., F., Lysebo, R. (2011). Comparison of electric power and propulsion plants for LNG carriers with different propulsion systems. ABB AS Oslo, Norway. http://www05.abb.com/global/scot. (Accessed December 2012).

[10]   Chang, D., Rhee, T., Nam, K., Chang, K., Lee, D., Jeon, S. (2008). A study on availability and safety of new propulsion systems for LNG carriers. *Reliability Engineering and System Safety* 93, 1877– 1885.

[11]   Zuancang, L., Yulong, Z., Qinming, T. (2009). Research on the selection of LNG carrier propulsion systems. *Pacific-Asia Conference on Knowledge Engineering and Software Engineering.* 19-20 Dec.

[12]   Wayne, W. S., Hodgson, M. (2006). The options and evaluation of propulsion of systems for the next generation of LNG carriers. 23rd World Gas Conference, Amsterdam.

[13]   Anderson, T. N., Ehrhardt, M. E., Foglesong, R. E., Jones, D., Richardson, A., Bolton, T. (2009). Shipboard Reliquefaction for Large LNG Carriers. Proceedings of the 1st

Annual Gas Processing Symposium H. Alfadala, G.V. Rex Reklaitis and M.M. El-Halwagi. Elsevier.

[14] Shin, Y., Lee, Y., L. (2009). Design of a boil-off natural gas reliquefaction control system for LNG carriers. *Applied Energy*. 86, 37–44.

[15] Sayyaadi, H., Babaelahi, M. (2010). Exergetic Optimization of a Refrigeration Cycle for Re-Liquefaction of LNG Boil-Off Gas. *International Journal of Thermodynamics*, 13, 4, 127-133.

In: Ships and Shipbuilding
Editor: José A. Orosa

ISBN: 978-1-62618-787-0
© 2013 Nova Science Publishers, Inc.

Chapter 3

REGRESSION METHODS APPLIED TO A GAIN SCHEDULING PID CONTROLLER TO GUARANTEE THE AUTOMATIC STEERING OF SHIPS

Héctor Quintián [*]*José Luis Calvo-Rolle* [†]
José Luis Casteleiro-Roca [‡]*and José Antonio Orosa García* [§]
University of A Coruña

Abstract

The present chapter shows the model designed to guarantee the automatic steering of a ship. The proposal is based on a typical PID (Proporcional-Integral-Derivative) controller, tuning in closed loop with relay-feedback method. The system is a typical non-linear case of study that requires a hard work to solve it. First, best tuning parameters of PID controller are obtained for the operation range of the ship. Then with the dataset, several intelligent methods of regression have been tested, to achieve the final solution. The best regression model is chosen to obtain the PID controller parameters under all possible working points.

1. Introduction

In the control engineering field, it is necessary to work in a continuous way to achieve new methods of regulation, in order to improve that already exists, or to find better alternatives, for example [Nunes et al., 2007, Begovich et al., 2007]. This dizzying demand of applications in control, is due to the wide range of possibilities developed until this moment.

Despite the great number of research and technical contributions, it has been impossible up to now that relatively popular techniques, such as the "traditional" PID control have fallen into disuse. Since the discovery of such regulators by Nicholas Minorsky in the field of automatic steering of ships [Mindell, 2002, Bennett, 1984] in 1922, many studies on

---

[*]E-mail address: hector.quintian@udc.es

[†]E-mail address: jlcalvo@udc.es

[‡]E-mail address: jose.luis.casteleiro@udc.es

[§]E-mail address: jaorosa@udc.es

this controller have been made. It should be noted that there are numerous usual control techniques for the process in any field, where innovations were introduced, for example the inclusion of artificial intelligence in this area, [Moradi, 2003]. But it is well-known, that the great majority of its implementation use PID controllers, raising the utilization rate up to 90% [Åström and Hägglund, 1995, Li et al., 1998]. The use of this type of controller remains very common for different reasons, such as toughness, reliability, simplicity, error tolerance, and so on.

The vast majority of real systems are not linear. This occurs notably in the steering of ships, from which long time ago, emerging models as the first or second order Nomoto model [Nomoto and Taguchi, 1957], the Norrbin model [Norrbin, 1963] or the Bech model [Bech and Smith, 1969]. Nowadays this topic remains a case of study [Bhattacharyya and Haddara, 2006, Casado et al., 2007], because according to their working point, certain specifications will be required to be equal in all areas of operation. Thus, different values of the regulator parameters will be needed in each of these areas. Taking this into account, self and adaptive PID regulators [Moradi, 2003, Åström and Wittenmark, 1994, Åström and Tore, 2001, Camacho and Bordons, 1997] are a good solution to reduce this problem. Though it should be noted that its implementation is quite difficult, expensive and closely linked to the type of process which purports to regulate, being sometimes difficult to establish a general theory in this type of PID controllers.

To solve this problem the well-known Gain Scheduling method can be applied, which is easier to implement, and with which highly satisfactory results are obtained. The concept of Gain Scheduling arises at the beginning of the 90´s [Rugh, 1991], and it is considered as part of the family of adaptive controllers [Åström and Wittenmark, 1994]. The principle of this methodology is to divide a non-linear system in several regions in which its functioning is linear. Thus, the parameters of the controller are obtained to allow having similar specifications around the operating range of the plant.

To implement the Gain Scheduling, first it is necessary to choose the significant variables of the system according to which it is going to be defined the working point. Then it is necessary to choose operating points along the entire range of functioning of the plant. There is no systematic procedure for these tasks. Often the first step is to take those variables that could be measured easily. The second step is more complicated because of the points that have to be selected. The system can be stable for the parameters deducted for the controller, but it does not have to be stable between the selected points. This problem has no simple solution, and when it exists, there is usually particularized. That is why this subject has been studied by researchers, see for example [Clement and Duc, 2001, Lu et al., 1991, Hiramoto, 2007].

A way to solve the problem is using regression technics to implement the Gain Scheduling method. The aim of the present work is to create a regression model to choose the best combination of PID controller parameters depending of the working point of the process. Three technics for regression have been taken into account: Support Vector Regression (SVR), Artificial Neural Networks (ANN), and Polynomial Regression.

This document is structured starting with a brief introduction of the topology of PID controller, which it is used after to show an explanation of the three regression methods proposed. Then, a description of its application to the model of a ship is presented. The model of the ship, to carry out the steering control with the proposed methodology, is a non-linear system, and it is applicable in different steering models of existing ships regardless of the complexity. Finally, the study ends with the validation of the method, making simulations under different operation points.

2. The PID Controler

There are multiple representation forms of PID controller, but perhaps the most widespread and studied is the one given by the equation 1.

$$u(t) = K\left[e(t) + \frac{1}{T_i}\int e(t)dt + T_d\frac{de(t)}{dt}\right] \quad (1)$$

Where u is the control variable and e is the control error given by $e = YSP - y$ (difference between the reference specified by the input and the output measured in the process). Thus, the control variable is a sum of three terms: the term P, which is proportional to the error, the term I, which is proportional to the integral error, and the term D, which is proportional to the derivative of error. The controller parameters are: the proportional gain K, the integral time T_i and the derivative time T_d.

There are multiple ways for the representation of a PID controller, but to implement the PID controller used and defined in the formula above, and more commonly known as the standard format [Åström and Hägglund, 1995, Li et al., 1998], shown in representation bloc, it is shown in figure 1.

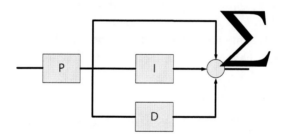

Figure 1. PID controller in standard format

There are a lot of processes in industries whose normal function is not adequate for certain applications. The problem is often solved by using the PID controller, by which the system is going to obtain certain specifications in the process control leading it to optimal settings for the certain process. The adjustment of this controller is carried out by varying the proportional gain, and the integral and derivative times commented before.

3. Adjustment Methods of the Parameters Controller with Gain Scheduling

On many occasions this method is known as the process dynamic changes with the process operating conditions. One reason for the changes in the dynamic can be caused, for example, by the well-known nonlinearities of processes. Then it is possible to modify the control parameters, monitoring their operational conditions and establishing appropriate rules. The methodology starts with the application of Gain Scheduling, analyzing the functioning of the plant in different working points, and then implementing different Regression Methods to program the parameter of the controller. Using this method is possible to obtain certain specifications which remain, in the possible extent, constant throughout the whole range of operation of the process. This idea can be schematically represented as shown in figure 2.

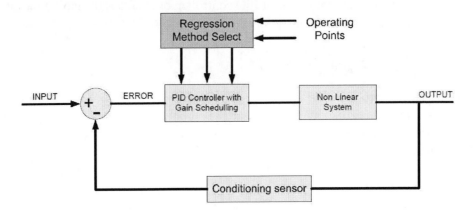

Figure 2. Gain Scheduling control schematic

The Gain Scheduling method can be considered as a non-linear feedback of a special type, because it is used a linear controller whose parameters are modified depending on the operation conditions. The idea is simple, but its implementation is not easy to carry out, with the exception of the computer controlled systems. As it is shown in figure 2, the working point of the process, change the parameters of the controller by the Regression Method selected.

4. Implementation with Regression Methods Instead of the Knowledge Base Tuning Rules

The replacement of the rules that define the gains of PID controller based on the working point system, is raised by a regression method whose inputs are the operation conditions of the plant, and as outputs the parameters of PID controller (K, T_i and T_d). The methods used in this research are the Support Vector Regression (SVR), Artificial Neural Networks (ANN), and Polynomial Regression.

A brief description of each method is presented in the next subsections.

4.1. Support Vector Regression (SVR)

Support Vector Regression (SVR) is a modification of the algorithm of the Support Vector Machines (SVM) for classification. In SVR the basic idea is to map the data into a high-dimensional feature space F via a nonlinear mapping and to do linear regression in this space.

Mathematically, if we have a given training data $\{(x_1, y_1), ..., (x_l, y_l)\}$ with $X \in \mathfrak{R}^n$ and $Y \in \mathfrak{R}$, the linear SVR algorithm tries to find the function:

$$f(x) = \langle w, z \rangle + b \tag{2}$$

Where $\langle \cdot, \cdot \rangle$ denotes the dot product in X, minimizing:

$$\frac{1}{2} \|w\|^2 + C \sum_{i=1}^{l} Q(y_i - f(x_i)) \tag{3}$$

The constant $C > 0$ determines the trade-off between the flatness of f and the amount up to which deviations larger than ε are tolerated. $Q(x) = max\{0, |x| - \varepsilon\}$ corresponds to Vapnik's $\varepsilon - insensitive$ loss function [Vapnik, 1995] (figure 3), which does not penalize errors less than $\varepsilon \geq 0$. If we take into account the case of nonlinear regression and reformulate, the optimization of the problem transforms it into the minimization of:

$$\frac{1}{2} \sum_{i,j=1}^{l} (\alpha_i - \alpha_i^*)(\alpha_j - \alpha_j^*) k(x_i, x_j) + \varepsilon \sum_{i=1}^{l} (\alpha_i - \alpha_i^*) - \sum_{i=1}^{l} y_i(\alpha_i - \alpha_i^*) \tag{4}$$

subject to:

$$\sum_{i=1}^{l} (\alpha_i - \alpha_i^*) = 0 \ and \ \alpha_i, \alpha_i^* \in [0, C] \tag{5}$$

Where the α_i, α_i^* are Lagrange multipliers and $k(x_i, x_j)$ is a kernel function, defining the future space in which the optimal solution of the problem will be computed in order to handle nonlinear problems. To estimate a new point the following function is used:

$$f(x) = \sum_{i=1}^{N} (\alpha_{s_i} - \alpha_{s_i}^*) + b \tag{6}$$

Where s_i are the indices of the data points for which either α_{s_i}, or $\alpha_{s_i}^*$ is non-zero. Those points are the support vectors (figure 3).

Least Square Support Vector Machine (LS-SVM).

Least Square formulation of SVM, are called LS-SVM. The approximation of the solution is obtained by solving a system of linear equations, and it is comparable to SVM in terms of generalization performance [Ye and Xiong, 2007]. The application of LS-SVM to regression is known as LS-SVR (Least Square Support Vector Regression). In LS-SVR, the $\varepsilon - insensitive$ loss function is replaced by a classical squared loss function, which constructs the Lagrangian by solving the linear Karush-Kuhn-Tucker (KKT) system:

$$\begin{bmatrix} 0 & I_n^T \\ I_n & K + \gamma_{-1}I \end{bmatrix} \begin{bmatrix} b_0 \\ b \end{bmatrix} = \begin{bmatrix} 0 \\ y \end{bmatrix} \tag{7}$$

Where I_n is a $[nx1]$ vector of ones, T means transpose of a matrix or vector, γ a weight vector, b regression vector and b_0 is the model offset.

In LS-SVR, only 2 parameters (γ, σ) are needed. Where σ is the width of the used kernel [Yankun et al., 2007].

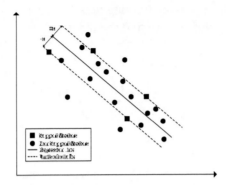

Figure 3. SVR linear regression with ε − *insensitive* loss function

4.2. Artificial Neural Networks (ANN)

The Artificial Neural Networks are computational algorithms based on the functioning of the human brain. Once of the most used ANN is the MLP (Multilayer Perceptron) [Bishop, 2006]. The MLP is composed by one input layer, one or more hidden layers and one output layer (see figure 4), all of them made of neurons and with pondered connections between neurons of each layer. Applying the *Theorem of Universal Approximation* [Hornik et al., 1989], it can be demonstrated that only one hidden layer is needed to model a nonlinear projection between input and output layer.

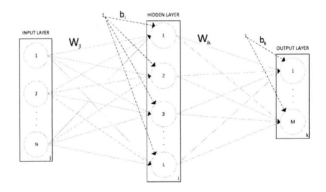

Figure 4. Architecture of Multilayer Perceptron with 1 hidden layer

A MLP with one hidden layer, can be written mathematically as follows:

$$y_k^p = F_k \left(\sum_{i=1}^{L} w_{ik} \cdot F_i \left(\sum_{j=1}^{N} w_{ji} \cdot x_j^p + b_i \right) + b_k \right) \tag{8}$$

Where:
$F_k \rightarrow$ Activation function of neurons of the output layer.
$w_{ik} \rightarrow$ Weight vector of connections from neurons of hidden layer to neurons of output layer.
$b_k \rightarrow$ Bias of neurons of the output layer.

$k \rightarrow$ Number of neurons of the output layer.
$F_i \rightarrow$ Activation function of neurons of the hidden layer.
$w_{ji} \rightarrow$ Weight vector of connections from neurons of input layer to neurons of hidden layer.
$b_i \rightarrow$ Bias of neurons of the hidden layer.
$i \rightarrow$ Number of neurons of the hidden layer.
$x_j^p \rightarrow$ p-th input pattern.
$j \rightarrow$ Number of neurons of the input layer (equals to dimension of the input data).
$y_k^p \rightarrow$ Predicted output for the p-th input pattern.

4.3. Polynomial Regression

Generally, a polynomial regression model may also be defined as a linear summation of basic functions:

$$F(x) = \sum_{i=1}^{k} a_i f_i(x) \tag{9}$$

Where k is the number of basic functions (equal to the number of model's parameters); $f_i(x)(i = 1, 2, ..., k)$ is a predefined polynomial basic function. The number of the basic functions in polynomial model of degree p is:

$$m = \prod_{i=1}^{p} (1 + d/i) \tag{10}$$

The estimation of model's parameters is made based on the training data, typically using the Ordinary Least-Squares (OLS) method, minimizing:

$$a = arg\, min_a \sum_{i=1}^{n} \left(y_{(i)} - F\left(x_{(i)} \right) \right)^2 \tag{11}$$

Where $x_{(i)}$ is the vector of input variables' values of the i_{th} data point and $y_{(i)}$ is the output value of that point. In this polynomial method as well as all the other regression modeling methods, the systems of linear equations in OLS are solved using Gaussian elimination and back substitution.

5. Nomoto Model of Ship-Steering Process

To analyze a ship's dynamics as a Nomoto model it is necessary to define a coordinate system as indicate in figure 5.

We consider V as the total velocity, u and v the x and y components of the velocity, and r the angular velocity of the ship. In normal steering the ship makes small deviations from straight-line course. The natural state variables are the sway velocity v, the turning rate r, and the heading ψ. The equations 12 are then obtained, taking into account that u is the

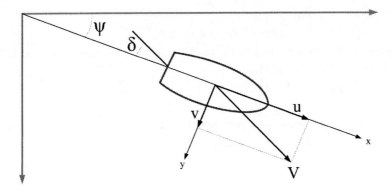

Figure 5. Coordinates and notation used to describe the equations

constant forward velocity, l the length of the ship and a and b are parameters of the model ship.

$$\begin{aligned}\frac{dv}{dt} &= \frac{u}{l}a_{11} + ua_{12}r + \frac{u^2}{l}b_1\delta \\ \frac{dr}{dt} &= \frac{u}{l^2}a_{21}v + \frac{u}{l}a_{22}r + \frac{u^2}{l^2}b_2\delta \\ \frac{d\psi}{dt} &= r\end{aligned} \qquad (12)$$

From equations 12 is determinated the transfer function from the rudder angle to the heading in the equation 13.

$$G(s) = \frac{K(1+sT_3)}{s(1+sT_1)(1+sT_2)} \qquad (13)$$

Where:
$K = K_0 u/l$
$T_i = T_{i0} l/u \qquad i = 1,2,3$

The parameters K_0 and T_{i0} are parameters of the ship model. In many cases the model can be simplified to equation 14.

$$G(s) = \frac{b}{s(s+a)} \qquad (14)$$

Where:
$b = b_0 \left(\frac{u}{l}\right)^2 = b_2 \left(\frac{u}{l}\right)^2$
$a = a_0 \left(\frac{u}{l}\right)$

This model is called the first order Nomoto model of a ship. Its gain b can be expressed approximately as expression 15.

$$b = c \left(\frac{u}{l}\right)^2 \left(\frac{Al}{D}\right) \qquad (15)$$

Where D is the displacement (in m^3), A is the rudder area (in m^2) and c is a parameter whose empirical value is approximately 0.5. The parameter a will depend on trim, speed and loading and its sing may change with the operating conditions.

6. System Used to Verify the Proposed Method

The method proposed in this document for the automatic steering of ships is going to be applied to a freighter of 161 meters of length. The displacement of the freighter will range from 8000 m^3 in the vacuum until the 20000 m^3 in full load. The velocity at which the ship will be able to navigate, will be more than 2 meters per second, for which is perfectly valid the model used, to 8 meters per second at maximum velocity. It is necessary to specify that the servo-rudder operates at a speed of 4 meters per second limited to ± 30 degrees, according to the description of the model shown in figure 6.

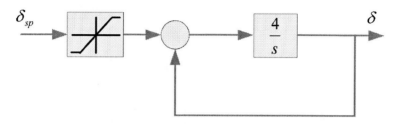

Figure 6. Servo-rudder blocks diagram

The transfer function of the freighter is indicated in the expression 16, in which was only given the value to c that is 0.5.

$$G(s) = \frac{0.5 \left(\frac{u}{l}\right)^2 \frac{Al}{D}}{s\left(s + a_0 \frac{u}{l}\right)} \tag{16}$$

It will be replaced the fixed values commented, where it should be pointed out for the case of the freighter, that the parameter a_0 has a value of 0.19 and the variable will be replaced in each case as necessary.

To carry out the simulations and to achieve the desired data for implementing the proposed model, it is used *Matlab/Simulink*®. For which it is edited, firstly, the following control scheme based on the descriptions made previously, shown in figure 7.

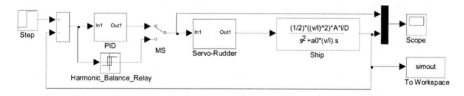

Figure 7. Control scheme in *Simulink*® format

The PID block contains the scheme shown in figure 8, whose structure is similar to the figure 1 of the document, but in which it has been noted, for a better approximation to the real system, the congestion at the output of the block adder.

The servo-rudder block has inside the diagram of figure 9, which is the diagram of the Servo (figure 6) in *Simulink*® format.

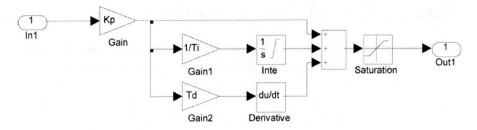

Figure 8. PID block in Simulink format

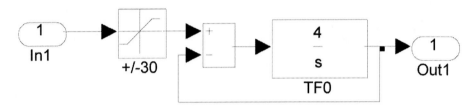

Figure 9. Servo-rudder in Simulink format

6.1. System Operation Conditions

The system operation conditions are infinite so that certain values have to be chosen. It makes no sense to obtain parameters for multiple cases, so it is necessary to make a coherent estimation to achieve good results. One approach is to choose a reasonable amount of equidistant values and to observe the changes of the parameters for each case. If there are substantial changes from one value to another then an opportunity of taking new intermediate values between them will be provided.

There are own characteristics of the ship that will not vary, for example the area of the rudder. In this case, the only terms that will define the operation conditions or the gains adjustment rules of the controller, are the displacement and the velocity of the ship. It is necessary to highlight that the displacement is not going to be a property in constant changing,as can be the velocity, even if it is in a slow way. Taking into account the previous comments and the range of values that can take each of the two parameters (displacement and velocity), the table 1 of possible conditions is established.

6.2. Obtaining the Controller Parameters for Each Operation Status

In the stage of obtaining the parameters of different working points; in the control implementation of the virtual controller, instead of a PID controller in parallel, could be used an hysteresis block. This is an attempt to obtain the controller parameters using the Relay-Feedback method and it is discussed in a summarized way.

6.2.1. Relay-Feedback Method

This is an alternative way to the chain closed method of Ziegler-Nichols [Zhuang and Atherton, 1991, Cominos and Munro, 2002, Åström et al., 1998, Hägglund and Åström, 2002, Ziegler and Nichols, 1993], for the empirical location of the critical gain (K_c) and

Table 1. Working points selected

D	u
8000	2
	4
	6
	8
12000	2
	4
	6
	8
16000	2
	4
	6
	8
20000	2
	4
	6
	8

the period of sustained oscillation (T_c) of the system. It uses the method of Relay (Relay-Feedback) developed by Åström and Hägglund [Åström and Wittenmark, 1994, Åström and Hägglund, 1984], which consists in leading the system to the oscillation state by the addition of a relay as it is shown in figure 10.

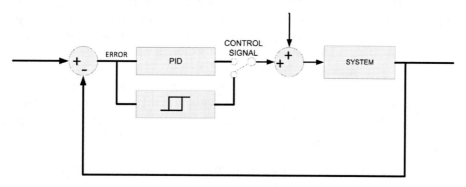

Figure 10. Application scheme of Relay-Feedback

This oscillation taken from the system has a period with approximately the same value as the period of sustained oscillation T_c (critical period). In the experiment it is recommended to use a relay with hysteresis, with the characteristics shown in figure 11, with an amplitude d and a width of the hysteresis window h.

After the assembly is done, the procedure to get the parameters mentioned, is the following:

1. Leading the process to a steady state, with the system regulated by the PID controller and with some parameters that let us achieve that status. Notes of the control signal

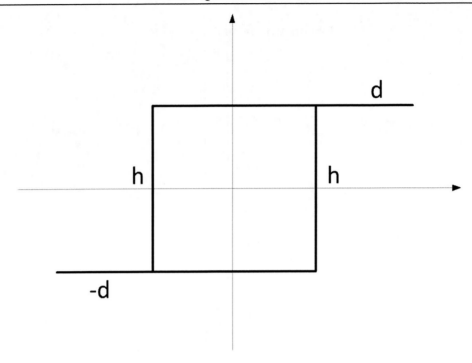

Figure 11. Hysteresis for Relay-Feedback

values and the output of the process in those conditions are taken.

2. Then the control is finished with the relay, instead of with the PID controller. As a set point it is given the value of the output of the process in the previous step. This value is introduced in the input shown in figure 10 as the $Offset$, which is necessary to put the process in steady state.

3. The process is put into operation with the indications made in the previous paragraph, and it is expected to become regular in the output (in practice it can be considered to have reached this state when the maximum value of the output repeats the same value for at least two consecutive periods).

4. It will be noted down the two parameters shown in figure 12, where T_c is the sustained oscillation period.

5. The critical gain of the process is determined by the expression 17.

$$K_c = \frac{4d}{\pi\sqrt{a^2 - h^2}} \qquad (17)$$

The Relay-Feedback has the advantage that adjustments can be made on the set point and it can be carried out at any time. However, the main problem is that during the tuning process, the set point can be exceeded on several occasions and there could be cases in which this is inadvisable because of the damage caused during the process.

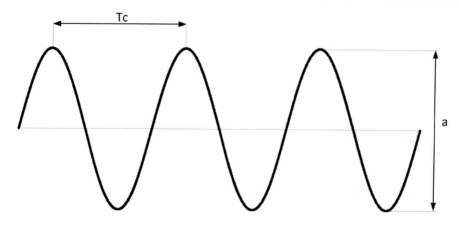

Figure 12. Parameters to read in the sustained oscillation (critic period)

6.2.2. Obtaining the Parameters T_c and K_c

In the particular case of this paper, there is no need to implement the hysteresis mentioned in the explanation of Relay-Feedback with a window, because the case of this article works with a relatively slow system. A simple comparator as the one shown in figure 13 is sufficient.

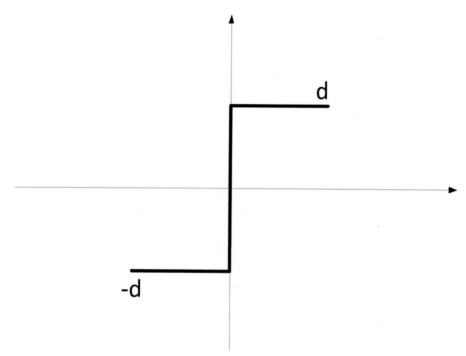

Figure 13. Hysteresis for the plant tested

Logically the value of h is zero and the value of d is 0.5. It is established as set point a value of 0.5, and the offset for this case is not necessary because it would be zero. Under these conditions the system becomes operational, and the result obtained is shown in the figure 14.

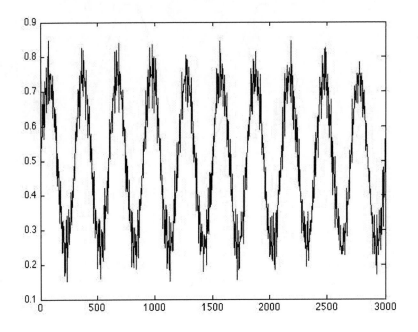

Figure 14. Result of application of Relay-Feedback in a working point

It is necessary to pay attention to the final zone, where the oscillation is now stabilized and periodic, and with the expressions commented before for the Relay-Feedback method, the extracted parameters are T_c and K_c.

6.2.3. Obtaining the PID Controller Initial Parameters

With the parameters that have been obtained in the previous paragraph, it is possible to calculate the controller parameters applying direct formulas, achieving the three terms of the regulator. In this case it will be necessary to obtain them for a criterion of changes in the load (for load disturbances rejection). Taking this into account, the expressions to tune controllers in closed loop of Ziegler-Nichols can be applied. They are the pioneer formulas for obtaining controller parameters, and they are good with changes in the load. The specification, that it is trying to obtain, is a list of overshoot of a quarter decay ratio, which means that in the face of the input of a disturbance, the successive overtopping of reference is four times lower than the previous one(damping factor of 1/4). Such expressions are shown in Table 2.

Table 2. Ziegler-Nichols formulas for closed chain

K	$=0.6 \times K_c$
T_i	$=0.5 \times T_c$
T_d	$=0.125 \times T_c$

6.2.4. Fine Tuning of the Controller

The parameters obtained in the preceding paragraph would be necessary subjected to a fine-tuning, because the results reached are not suitable.

In this most delicate task of adjustment it is necessary to indicate that it should not saturate the output controller at any time. It is necessary to reach a compromise, because an excess of proportional gain causes a fast response in the output with little overshoot, which apparently is ideal. But under these conditions the servo is constantly fluctuating, which will cause it a deterioration in a short period of time. As a conclusion, it is necessary to search gradual outputs, without saturation or sudden changes such as the case shown in figure 15.

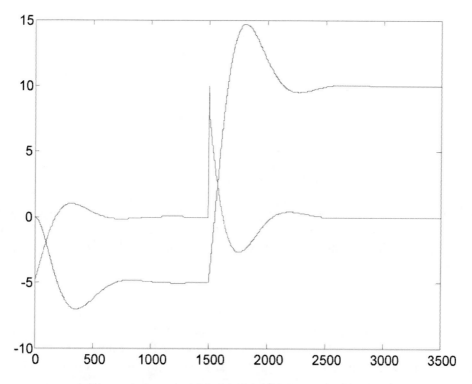

Figure 15. Example of steering and control signal to the rudder

It is necessary to indicate that for this model the method of adjusting parameters Ziegler-Nichols is not the ideal, since the initial values of the parameters do not give good results, differing greatly from those achieved after fine-tuning.

6.2.5. Parameters Obtained for Each Case

Taking the previous comments into account, we obtain the controller parameters fine-tuned for each of the cases discussed before. We seek a criterion of minimum overshoot and maximum speed for the restrictions presented in the preceding paragraphs. In this way we reach the parameters of the table 3.

Table 3. Controller parameters obtained for each rule

D	u	K	T_i	T_d
8000	2	3	350	90
	4	1	300	80
	6	1	350	80
	8	1	350	80
12000	2	4	350	80
	4	3	300	90
	6	1	350	95
	8	1	350	85
16000	2	6	400	90
	4	3	300	90
	6	1	350	90
	8	1	300	85
20000	2	6	450	90
	4	4	350	90
	6	1	350	90
	8	1	300	90

6.3. Implementation of the Regression Methods

Several tuning processes for each method have been made to ensure that each resulting model was the best for the corresponding method. In the next subsections it is explained, without entering in deep computational analysis, the tuning of each method.

6.3.1. Support Vector Regression (SVR)

LS-SVM [De Brabanter et al., 2010] Matlab toolbox has been used. In this toolbox, the tuning of the parameters (γ, σ) (expression 7) is conducted in two steps. First, a state-of-the-art global optimization technique, Coupled Simulated Annealing (CSA) [Xavier de Souza et al., 2010], determines suitable parameters according to some criterion. Second, these parameters are then given to a second optimization procedure (simplex or grid search) to perform a fine-tuning step.

The optimal parameters obtained by the previous method are: $\gamma = 2.1515e^7$, $\sigma2 = 1.99477$.

6.3.2. Artificial Neural Networks (ANN)

A neural network, type MLP (Multi Layer Perceptron), has been designed for scheduling each one of the controller constants K, T_i and T_d. The neuronal network has an intermediate layer with 5 neurons for K and 6 for T_i and T_d. This structure has been adopted after many tests with different numbers of neurons in the middle layer (tests were made from 4 to 9 neurons in the middle layer) for each of the neural networks. The activation functions of neurons in the middle layer are a kind of hyperbolic tangent, except in the output layer, in which one neuron has a linear function.

Once this configuration is selected, it is shown the different characteristics of the training carried out with backpropagation learning. It has been made the training of K, T_i and T_d at 531, 705 and 686 respectively epochs, with an average error at the end of the training less than 1%. The artificial neural networks have been trained off-line, although the checking of its proper operation has been performed on line.

6.3.3. Polynomial Regression

Several tests were made to achieve the optimal values of the different parameters. At first, it was necessary to discover the optimal degree of the Polynomial Regression for each constant of the PID. At the end of the tests, it was found that the best results were achieved with $d = 2$ to calculate the gain, and $d = 1$ to calculate the time constants.

In the implementation of the regression, it would be necessary to define two different inputs to the polynomial, one to calculate the K, one other to calculate the timing constants (T_i and T_d). The average error at the end of the training was around 2.8%.

7. System Assembly and Verification of Results

The system diagram is implemented in $Simulink^®$ (figure 16). If we compare this system diagram with the model used for obtaining the controller parameters in the different representative points of work, it we can observe that the relay block has been eliminated, and the PID controller has been replaced for two blocks, with the name of 'Gain Scheduling Regression Method' and 'PID'.

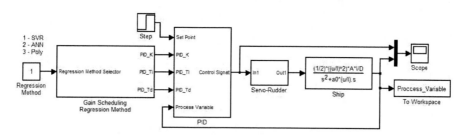

Figure 16. System on Simulink format

This program allows us to compare the function of the proposed methods with the same simulate input signal.

It is necessary to remember that the real implementation would only need to use one on the methods, not all of them. However, if the results of the simulations show that the best method is and hybrid of all of them; they could be implemented together, and the working point would select the best one according to the simulations.

All the methods have the same two inputs to determine the working point of the system: the Displacement (D) and the Velocity (u) of the ship. The internal diagram blocks of the Regression Gain Scheduling selector is shown in figure 17 in the format implemented in $Simulink^®$.

80   Héctor Quintián-Pardo et al.

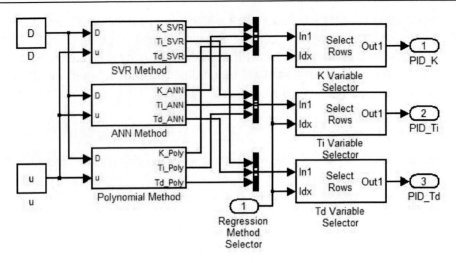

Figure 17. Regression Method selector

In this way, the implemented controller will choose the most appropriate parameters for the area of functioning. It should be noted that more points could be obtained to train the Regression Methods, but it would be more expensive. Furthermore, the best results are achieved with the Neural Network itself following the tendency of data, already interpolated properly between them, showing one of the advantages of its use. To validate the model created it is resorted to its simulation with different values of the parameters, on which depend the velocity model and displacement.

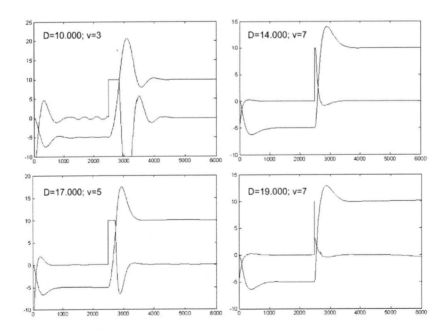

Figure 18. Model response to different operating conditions - ANN

Different tests at multiple working points, have been made and in figure 18 are shown four representative examples for the best Method. In the test, it is made a steering to $-5°$ and once stabilized to $+10°$. The results are satisfactory and similar in all cases with the only difference of the respond velocity due to the different velocities of the ship. The differences between the Regression Methods are basically the process time and the running expensive resources.

It should be stressed that at low velocities, to get an adequate response of the steering (similar to one with the entire range), it is necessary to saturate the output of the controller. This action is unwanted at the time of fine-tuning, but necessary to maintain the specifications within a range of values. After the results of the implemented system, it is emphasized the satisfactory operation of it. The desired results of uniformity are achieved in the operation, regardless the conditions, from which depends the model of the ship.

8. Conclusion

Obviously, in non-linear systems, such as the case of the steering of a ship studied in this document, working across a wide range of operation is very complicated. If it is possible to divide in areas with a linear behavior, and in which the control is also feasible using a type PID controller, the option of using the method proposed in this paper is a solution to take into account. As an alternative to the different types of auto-tuning PIDÕs, one of the easiest solutions is the one developed in this article. It is necessary to emphasize that it is not an easy solution to adopt, especially with continuous controllers, but with the addition of programmable control devices, this labour becomes comparatively simple. The use of PID controllers working with Gain Scheduling, has the problems of taking the significant points, the interpolation between them and also can happen that the system is stable at selected points but not between them. With the use of artificial neural networks, all these drawbacks are softened to a large extent, because all of them are solved with the application of artificial intelligence. This methodology provides a uniform response of the system throughout the whole operating range of the ship, regardless of the displacement, the velocity, or other parameters of which the model depends on.

References

[Bech and Smith, 1969] Bech, M. I. and Smith, L. W. (1969). Analogue simulation of ship manoeuvers. Technical report, Hydro and Aerodynamics Laboratory.

[Begovich et al., 2007] Begovich, O., Ruiz, V. M., Besancon, G., Aldana, C. I., and Georges, D. (2007). Predictive control with constraints of a multi-pool irrigation canal prototype. *Latin American applied research*, 37:177–185.

[Bennett, 1984] Bennett, S. (1984). Nicholas minorsky and the automatic steering of ships. *Control Systems Magazine, IEEE*, 4(4):10–15.

[Bhattacharyya and Haddara, 2006] Bhattacharyya, S. K. and Haddara, M. R. (2006). Parametric identification for nonlinear ship maneuvering. *Journal of Ship Research*, 50(3):197–207.

[Bishop, 2006] Bishop, C. M. (2006). *Pattern recognition and machine learning*. Springer.

[Camacho and Bordons, 1997] Camacho, E. F. and Bordons, C. A. (1997). *Model Predictive Control in the Process Industry*. Springer-Verlag New York, Inc.

[Casado et al., 2007] Casado, M. H., Ferreiro, R., and Velasco, F. J. (2007). Identification of nonlinear ship model parameters based on the turning circle test. *Journal of Ship Research*, 51(2):174–181.

[Clement and Duc, 2001] Clement, B. and Duc, G. (2001). An interpolation method for gain-scheduling. In *Decision and Control, 2001. Proceedings of the 40th IEEE Conference on*, pages 1310–1315.

[Cominos and Munro, 2002] Cominos, P. and Munro, N. (2002). Pid controllers: recent tuning methods and design to specification. *Control Theory and Applications, IEE Proceedings*, 149(1):46–53.

[De Brabanter et al., 2010] De Brabanter, K., Karsmakers, P., Ojeda, F., Alzate, C., De Brabanter, J., Pelckmans, K., De Moor, B., Vandewalle, J., and Suykens, J. A. K. (2010). Ls-svmlab toolbox user's guide version 1.7. Technical report, MatWorks.

[Hägglund and Åström, 2002] Hägglund, T. and Åström, K. J. (2002). Revisiting the ziegler-nichols tuning rules for pi control. *Asian Journal of Control*, 4(4):364–380.

[Hiramoto, 2007] Hiramoto, K. (2007). Active gain scheduling: A collaborative control strategy between lpv plants and gain scheduling controllers. In *Control Applications, 2007. CCA 2007. IEEE International Conference on*, pages 385–390.

[Hornik et al., 1989] Hornik, K., Stinchcombre, M., and White, H. (1989). Multilayer feedforward networks are universal approximators. *Neural Networks 2*, pages 359–366.

[Li et al., 1998] Li, Y., Feng, W., Tan, K. C., Zhu, X. K., Guan, X., and Ang, K. H. (1998). Pideasy(tm) and automated generation of optimal pid controllers. In *Proc. Third Asia-Pacific Conference on Measurement and Control*, pages 29–33.

[Lu et al., 1991] Lu, W. M., Zhou, K., and Doyle, J. C. (1991). Stabilization of lft systems. In *Decision and Control, 1991., Proceedings of the 30th IEEE Conference on*, pages 1239–1244.

[Mindell, 2002] Mindell, D. A. (2002). *Between Human and Machine: Feedback, Control, and Computing Before Cybernetics*. Johns Hopkins studies in the history of technology. The Johns Hopkins University Press.

[Moradi, 2003] Moradi, M. H. (2003). New techniques for pid controller design. In *Control Applications, 2003. CCA 2003. Proceedings of 2003 IEEE Conference on*, volume 2, pages 903–908.

[Nomoto and Taguchi, 1957] Nomoto, K. and Taguchi, K. (1957). On the steering qualities of ships (2). *Journal of the Society of Naval Architects of Japan*, 101.

[Norrbin, 1963] Norrbin, N. H. (1963). On the design and analyses of the zig-zag test on bases of quasi to line frequency response. Technical report, The Swendish State Experimental Shipbuilding Tank (SSPA).

[Nunes et al., 2007] Nunes, G. C., Rodrigues Coelho, A. A., Rodrigues Sumar, R., and Goytia Mejía, R. I. (2007). A practical strategy for controlling flow oscillations in surge tanks. *Latin American applied research*, 37:195–200.

[Åström and Hägglund, 1984] Åström, K. J. and Hägglund, T. (1984). Automatic tuning of simple regulators with specifications on phase and amplitude margins. *Automatica*, 20(5):645–651.

[Åström and Hägglund, 1995] Åström, K. J. and Hägglund, T. (1995). *PID Controllers: Theory, Design, and Tuning, 2nd Edition.* ISA.

[Åström et al., 1998] Åström, K. J., Panagopoulos, H., and Hägglund, T. (1998). Design of pid controllers based on non-convex optimization. *Automatica*, 34(5):585–601.

[Åström and Tore, 2001] Åström, K. J. and Tore, H. (2001). The future of pid control. *Control Engineering Practice*, 9(11):1163–1175.

[Åström and Wittenmark, 1994] Åström, K. J. and Wittenmark, B. (1994). *Adaptive Control.* Addison-Wesley Longman Publishing Co., Inc.

[Rugh, 1991] Rugh, W. (1991). Analytical framework for gain scheduling. *Control Systems, IEEE*, 11(1):79–84.

[Vapnik, 1995] Vapnik, V. (1995). *The Nature of Statistical Learning Theory.* Springer.

[Xavier de Souza et al., 2010] Xavier de Souza, S., Suykens, J. A. K., Vandewalle, J., and Bolle, D. (2010). Coupled simulated annealing. *IEEE Transactions on Systems, Man and Cybernetics - Part B*, 40(2):320–335.

[Yankun et al., 2007] Yankun, L., Xueguang, S., and Wensheng, C. (2007). A consensus least support vector regression (ls-svr) for analysis of near-infrared spectra of plant samples. *Talanta 72*, pages 217–222.

[Ye and Xiong, 2007] Ye, J. and Xiong, T. (2007). Svm versus least squares svm. In *11th International Conference on Artificial Intelligence and Statistics (AISTATS)*, pages 640–647.

[Zhuang and Atherton, 1991] Zhuang, M. and Atherton, D. P. (1991). Tuning pid controllers with integral performance criteria. In *Control 1991. Control '91., International Conference on*, volume 1, pages 481–486.

[Ziegler and Nichols, 1993] Ziegler, J. G. and Nichols, N. B. (1993). Optimum settings for automatic controllers. *Journal of Dynamic Systems, Measurement, and Control*, 115(2B):220–222.

In: Ships and Shipbuilding
Editor: José A. Orosa

ISBN: 978-1-6218-787-0
© 2013 Nova Science Publishers, Inc.

Chapter 4

KNOWLEDGE MODEL APPROACH BASED IN RULES FOR TACAN AIR NAVIGATION SYSTEM

Xosé Manuel Vilar Martínez [*]
University of Coruña
Juan Aurelio Montero Sousa [†]
University of Coruña
José Luis Calvo-Rolle [‡]
University of Coruña
José Luis Casteleiro-Roca [§]
University of Coruña
José Antonio Orosa García [¶]
University of Coruña

Abstract

The TACAN System is one of the most popular air navigation systems. Since its invention, it has been developed deeply because of its wide implementation, both in civil and military fields. It is a system difficult to understand, and with the aim of making its implementation easier, this paper describes the Knowledge Model based in rules developed for the TACAN system operation. The purpose of this paper is to achieve a better understanding of the TACAN system and its proper operation. The knowledge model of the paper has been tested using simulation with satisfactory results.

---

[*]E-mail address: x.vilar@udc.es
[†]E-mail address: j.montero.sousa@udc.es
[‡]E-mail address: jlcalvo@udc.es
[§]E-mail address: jose.luis.casteleiro@udc.es
[¶]E-mail address: jaorosa@udc.es

1. Introduction

The TACAN System (tactical air navigation acronym system) is one of the most used methods in the aircraft navigation [Kayton and Fried, 1997]. Since the initial model, in 1949 [Wikipedia, 2011], a lot of contributions have been made by technicians and researchers like [Gao et al., 2011, Bin and Wei-kang, 2009, Hadeii et al., 2011, deFaymoreau, 1956, Shi et al., 1996a, Wood-Hi, 1994, Cheng et al., 1993, Shi et al., 1997, Shi et al., 1996b]. Despite these, it is necessary to remark that it is not a trivial method with an easy understanding [Wikipedia, 2011, Kayton and Fried, 1997]. In this sense, the technical staff need a deep knowledge and enough experience to work with it.

Rule based systems are models that use the knowledge of human experts [F. Hayes-Roth, 1983a, Cimino et al., 2012]. They extract rules of a method and they structure it in accordance with its performance [F. Hayes-Roth, 1983a]. With these methods, the implemented system emulates the expert knowledge in certain fields [F. Hayes-Roth, 1983a, F. Hayes-Roth, 1983b], and has been one of the most implemented methodologies in researching and in operational areas [F. Hayes-Roth, 1983b]. There are some examples of these models [Costa et al., 2012, Chen et al., 2012, Cimino et al., 2012, Cline et al., 2010, ElAlami, 2011, Ishibuchi et al., 1999, Cordón, 2011, Sanz et al., 2011, K. Khalili-Damghani, 2011, Chang et al., 2012]. For instance [Costa et al., 2012] combines rule based system and case based reasoning to provide support for the product design decisions. In [Chen et al., 2012] is showed a rule-based system, complementing a Risk Metrics Wealth Bench system, for portfolio optimization with nonlinear cash flows and constraints. In [Cline et al., 2010] is described a robust and general rule-based approach, with the aim of managing the situation awareness.

Due to the difficulty of TACAN air navigation system, it is necessary to obtain a new approach that makes its implementation easier. Therefore, a knowledge model approach based on rules is created, with the aim of helping the air navigation TACAN technical staff. The proposal can be used also with educational objectives.

This paper is organized in the following sections: it starts with the present introduction, then a brief historical explanation is made, then, the next section shows the description of the TACAN technique. After that, a Knowledge Model approach is presented, following by the testing of the model using simulation. Finally the conclusions and future works are presented.

2. Historical Background

Humans have always wanted to fly, trying to imitate birds. In 1903, the Wright brothers made the first motorized flight. It was a very short flight using a catapult to propel the aircraft [Digital, 2012].

After this first flight, the development of the aviation was extremely slow. During the first decade of the twentieth century, the airplanes were rather scarce and the flights were too short, trying always to keep visual contact with the take off point. The First World War widened the scope of aviation to military use, contributing in this way to the development of aeronautics.

Civil aviation started with the postal service, especially in countries with a vast expanse of territory such as the United States. The pilots were guided by fires on the ground, especially during night flights, and they used also flares and road maps. There were few airports and they were located very far away one from another. The aid provided from the ground was scant and the weather usually prevented the flights or caused accidents.

In 1927 lighthouses located in different parts of the U.S. and 17 low and medium frequency (LF / MF) transponders began to operate. In 1929 there were 89 radio beacons on the airways. The radio communications between ground and planes was generally only in one direction, using a code until 1930. Since that year, the pilots were able to send messages from the cockpit radiotelephone to ground stations, helping the pilots in the guidance [Civil, 2012].

The Post Office Department of the U.S. carried out experiments to install radio equipment in their planes. The Federal Telecommunication Laboratories also performed a work to develop radio-navigation systems [Sandretto, 1956].

After so much work and a lot of experiments, safer means of navigation for aircrafts were designed, and TACAN, appeared [Wikipedia, 2012a].

The U.S. Navy had firstly specific problems related to the orientation of two objects in motion, a plane and a boat. To solve this, early text experiences with the background of the TACAN were performed on the USS Mississippi and USS Krause [Sandretto, 1956].

TACAN is the name of the military version and VOR-TACAN is the civilian version [Wikipedia, 2012a]. TACAN and VOR-TACAN share common elements, but obviously the military version (TACAN) is restricted to this use [Wikipedia, 2012b].

3. TACAN description

The TACAN system provides range and bearing, using the relative position of the aircraft beacon with the magnetic north.

These two variables are displayed on the indicators of the aircraft after decoding the received information.

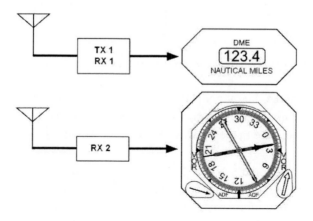

Figure 1. Information TACAN cockpit indicators

Figure 1 shows the information received in the cabin.

3.1. Range determination

The calculation of the range is based on the measuring of the signal time between the transmitter equipment and the aircraft transponder (ground equipment or ship). Once the time has been measured, the range is determined using the calculation expressed in the following equation 1, known as *DME* (Distance Measuring Equipment, figure 2), which is the range measured between the equipment. It is important to note that the transponder equipment delays the signal *R* in microseconds, and the signal varies depending on the work mode.

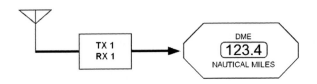

Figure 2. DME indicator

The table 1 shows the parameters for the different interrogation modes.
Where:
T / A-X are the ground-to-air channels in X mode, T / A-Y are the ground-to-air channels in Y mode, A / A-X are the air-to-air channels in X mode and A / A-Y are the air-to-air channels in mode Y.
N/A: Not Applicable, because it is a single pulse.

MODE	Interrogation pulse number	Separation between pulses−interrogation	Pulse number response	Separation between pulses−response	Delay R in μs
T/A-X	2	12 ± 0.1	2	12 ± 0.1	50 ± 0.1
T/A-Y	2	36 ± 0.1	2	30 ± 0.1	74 ± 0.1
A/A-X	2	12 ± 0.1	1	N/A	62 ± 0.1
A/A-Y	2	24 ± 0.1	1	N/A	74 ± 0.1

Table 1. Time separation between pulses interrogation and delay in the transponder

It is similar to determine the range using the sound with an echo.
In the case of electromagnetic waves which travel at the speed of light $3.108 \ m/s$, the expression for determining the range between equipment (DME) is shown in equation 1:
Where:
τ = Time measured from the query sending until that the response is received, units in μs.
R = Delay can be 50, 74, 62 μs depending on what mode you use (see table 1).
12.34667 = Round-trip time, in microseconds, in which the electromagnetic signal travels one nautical mile (see glossary, definition of nautical mile).

$$DME = \frac{\tau - R}{12.34667} = (Nautical \ Mile) \qquad (1)$$

3.2. Slant range and range referenced to ground

The range, measured by this method, is always the slant range. The range referenced to ground is achieved solving the Pythagorean triangle. The error will be greater with the increasing of height, the figure 3 shows this effect graphically. We should also keep in mind that, if the range is too big, the beacon should applied spherical geometry for exact range referred to ground. The range referred to ground must be taken into account for air navigation.

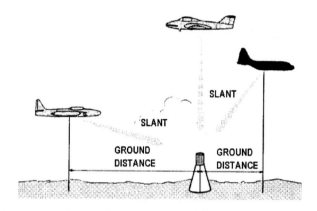

Figure 3. Slant range and range referenced to ground

If the height is measured in feet and the slant range in nautical miles, the expression for the range (in nautical miles) relative to Earth is shown in the following equation 2.
Where:
DRG = Referenced to ground range.
SD = Slant range.
H = Height in feet (ft).
6076.1 = Conversion factor from feet to Nautical Miles.
Operation units are obtained in Nautical Miles.

$$DRG = \sqrt{(SD)^2 - \left(\frac{H}{6076.1}\right)^2} \qquad (2)$$

3.3. Limitations of the determination method of the range

The method to determinate the range is simple, but it has a limitation. When the system try to attend a lot of DME requests, they aircrafts do not distinguish which is the answer for its own interrogation.

This limitation is solved in two ways. Firstly, the DME request is defined not only as a simple pulse, instead of that, the interrogation should consist in some pulses space in time. The aircraft includes a random times between each pulse. The same DME request sequence has very low probability of being repeated for difference aircrafts. Therefore, each aircraft could distinguish its answer and use it to determine the range (figure 4).

The times τ_1, τ_2 y τ_3 are random times, so the composition of the interrogation is unique.

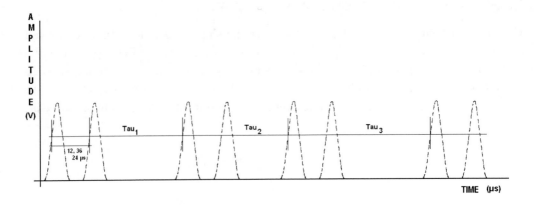

Figure 4. Coding interrogation separation between pulses

In the figure 5, it is shown a simple scheme of a DME transceiver.

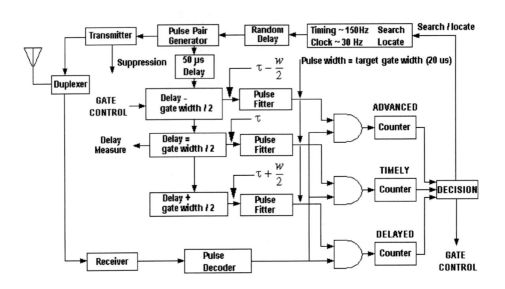

Figure 5. Simplified schematic of a DME transceiver

The circuit is composed by a transmitter, a receiver, and a duplexer between them. The function of the duplexer is to handle the derivation of the mass signal receiver. According to this, when the transmitter is transmitting, the receiver is not damaged.

Transmitter: the search mode is set for communication, with the transmission of a pulse sequence with a random delay. These delays are taken into account to recognize the pulse sequence.

Receptor: the received pulses are decoded and fed into a circuit that adjusts the pulse delay specified in the question. If the pulse falls within the expected time gate, it is recorded and a decision is made.

3.4. Bearing determination

The system calculates the delay or TACAN course. The aircraft should take that course to go to the beacon source. This course is achieved measuring the gap between the Main Reference Group (MRG) and the maximum amplitude in a modulated signal (figure 6).

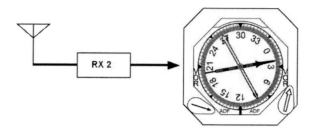

Figure 6. Heading indicator relative the magnetic north

Working mode	MRG pulses	ARG pulses	Separation time between pulses MRG μs	Separation time between pulses ARG μs
T/A-X	12 pares	6 pares	$12 \pm 0.1 - 30 \pm 0.1$	$12 \pm 0.1 - 24 \pm 0.1$
T/A-Y	13	13	30 ± 0.1	15 ± 0.1

Table 2. Reference Groups, Main and Auxiliary, function mode

The process, to determine the delay or direction in the receiver, is the following:

1- The receiver begins in search mode and performs the pulse count of the MRG, taking into account the work mode (table 2).

2- Once you locate the MRG pulses, the receiver starts a count to determine the separation between the MRG and the maximum amplitude of the signal.

3- A level detector circuit compares the amplitude modulation level with a reference level and when the maximum is detected, the account counter is informed, stopping its function immediately. Then the offset value of the group with respect to the main reference peak is obtained.

4- The conversion of time lag to angle is performed, using the equation 3, and then you get the delay or direction of the beacon.

$$Bearing = \frac{360 \cdot \delta}{66.6667} - 90 = (Magnetic\ North\ Course) \qquad (3)$$

Where:

Bearing = course, relative to magnetic north, that the aircraft should take to locate the beacon, (degrees).
δ = Signal offset of the MRG with the maximum amplitude modulated signal (measured in milliseconds *ms*).
360 = conversion factor to obtain the value measured in degrees.
66.6667 = pattern rotation period of the MRG (ms).
90 = correction for the delay or aircraft heading relative to magnetic north (degrees).

Some examples are presented now to illustrate what we explained before.

Example A

An aircraft receives a MRG signal with a lag of 0°.

Applying the equation 3, the result is obtained as −90° or 270°. Therefore the course to be taken by the aircraft A is 270° with respect to the magnetic north.

Example B

An aircraft B receives a MRG signal with a lag of 118°.

Applying the equation 3, the result is 28, so the course to be taken by the aircraft B is 28° relative to the magnetic north.

Example C

An aircraft C received a MRG signal with a lag of 245°. The aircraft is shown receiving the signal in the figure 7.

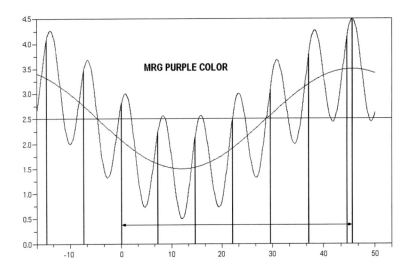

Delay between AM maximum and MBR is 45.370 ms.

Figure 7. Determination of bearing. Example C

Applying the equation 3, 155 is obtained as a result. Therefore the course to be taken by the aircraft is 155° with respect to the magnetic north.

Figure 8 shows the radial directions corresponding to the examples A, B and C.

3.5. Amplitude modulation of the TACAN system antenna

The aircrafts whose position with respect to the beacon are shown in the figure 8, receive the maximum signal amplitude, modulated with an offset, in function of the their position respecting the beacon. Therefore, the amplitude modulation of the position will vary as maximum rotates 360° in azimuth.

A TACAN antenna is shown in the figure 9. In this figure we can see the perspective of the antenna from the top of it.

Knowledge Model Approach Based in Rules for TACAN Air Navigation System   93

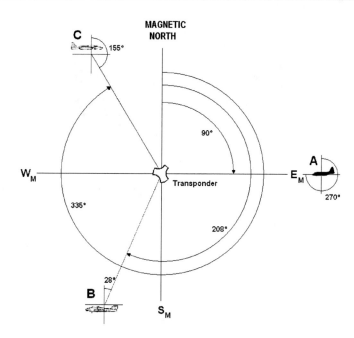

Figure 8. Radial from the beacon to the exercises A, B and C

In the central part of the antenna a radiating element is situated. This element is omnidirectional and transmits the power to the air. In a circumference closer to the radiant element, there is a reflector element that modifies the characteristics of the radiating element omni directionality. Then an example to clarify this idea is presented

Note that a point of light emits light with equal intensity in all directions (omnidirectional). If you introduce an object near the light, the object casts a shadow that will be larger as closer they are to the point of light. The shadow is changing the omni directionality of the light, whereby the light intensity is not the same at all the points of a concentric circle. If an observer moves throw the points of the circumference he will discover areas where the light intensity is higher than others.

The same situation is produced with the antenna-radiating element of TACAN. When a reflector element causes a "shadow" and modifies the radiation (the effect will be greater the closer you are to the radiating element). The antenna has another 9 elements (outside elements in figure 9) that modify the antenna radiation. The effect produced by the reflective elements (inside and outside) in the radiating element is a variation of the intensity radiated from the outside of the antenna (as an amplitude modulation).

From the point of view of the design of the antenna, it is intended that the antenna lobe have a cardioid-shaped radiation (figure 10).

Applying the previous paragraph, it is possible to obtain in the antenna radiation lobe. It represents the maximum of the signal in the East (by convention), and this is when the MRG is transmitted.

Each 40°, the system transmit an Auxiliar Reference Group (ARG) coinciding with each sub-maximum shows in figures 11 and 12.

The radiation lobe rotates, so, the aircraft receives the maximum at difference times

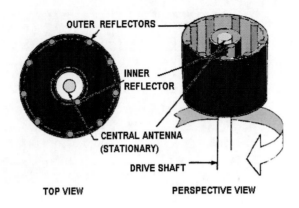

Figure 9. TACAN antenna, top view and perspective view

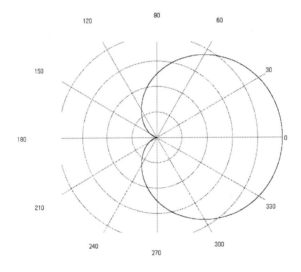

Figure 10. Cardioid on which performs amplitude modulation

depending of the situation of the transponder. Continuing with the examples, the radiation lobes of aircrafts B and C, are shown in figures 11 and 12.

The maximum radiation lobe aircraft A is in phase with the Main Reference Group.

The maximum radiation lobe aircraft B is 118° out of phase with the MRG, in figure 11 it is possible to appreciate this angle.

The maximum radiation lobe aircraft C is offset 245° with the MRG. Figure 12 shows this angle.

3.6. Precedence signals

Complete information transmitted by the TACAN order of precedence is:

1. Main Reference Group MRG is issued every 66.6*ms*.

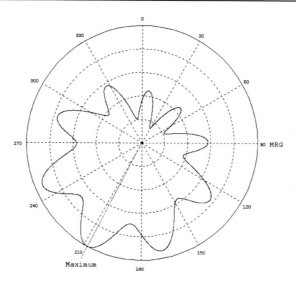

Figure 11. Radiation lobe of the antenna for example B

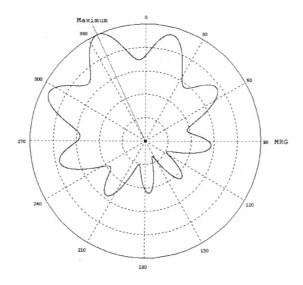

Figure 12. Radiation lobe of the antenna for example C

2. Auxiliar Reference Group, ARG. It airs every 7.4*ms*, except when you issue the MRG.

3. Group ID is issued every 30*s*.

4. Response interrogations (DME).

5. Squitter.

The figure 13 is a simplified diagram of a TACAN receiver.

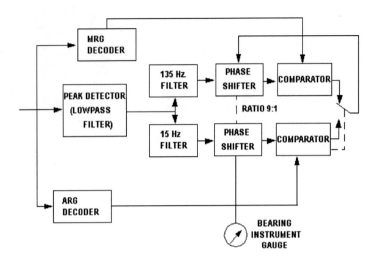

Figure 13. Simplified diagram of a receiver delay

First, the signal is decoded. Then, the receiver detects the MRG and the ARG. At the same time, the signal is analyzed with a low pass filter (peak detector), to find the maximum. Once the maximum is discovered, and the MRG is detected, the receiver could calculate the offset between them.

4. Knowledge model approach

The knowledge model begins by checking if the pulses transmitted have the correct specifications. The model identifies the MRG and ARG, and analyzes the timing between them. Finally it is checked the determination of the distance (DME). There is a model for X mode (figure 14) and another for Y mode (figure 15). Although the performance is similar for both models, the pulse spacing and number of pulses for each of them, are different.

5. Simulation of the model

Data has been collected from a TACAN system. The data were analyzed using decision methods and different parameters were achieved to ensure that the system works with the required specifications.

A programming example to check the pulse width is set forth below:

```
% Definition of variables:
% Data:   data matrix.
% B: array with data size.
% number_rows:  number of rows in the data matrix.
% n:   row counter.
% m:   index of the output vector.
```

Knowledge Model Approach Based in Rules for TACAN Air Navigation System

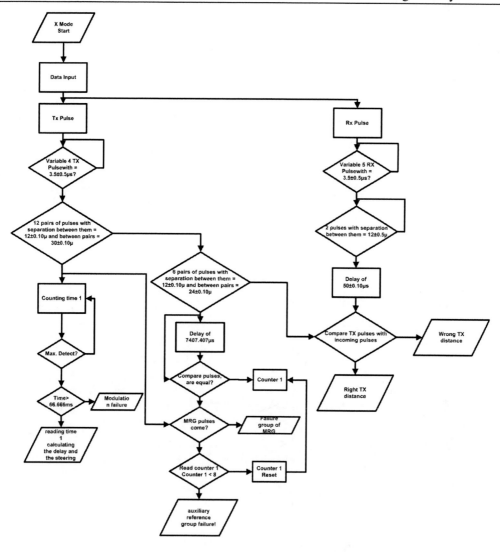

Figure 14. Knowledge model - X mode

```
% a0:  Voltage value at time t0.
% a1:  Voltage value at time t1.
% t0:  initial time value.
% t1:  time value end.
% m:   index vector pulse width.
% V: stores data column vector pulse width.
clear;
m=0; % initialize the variable m
Data = csvread ("Data_file.csv"); % Read Data_file.csv
B = size(Data); % Data matrix size
number_rows=B(1,1)-1; % number of rows of the matrix
for n=1:number_rows;
```

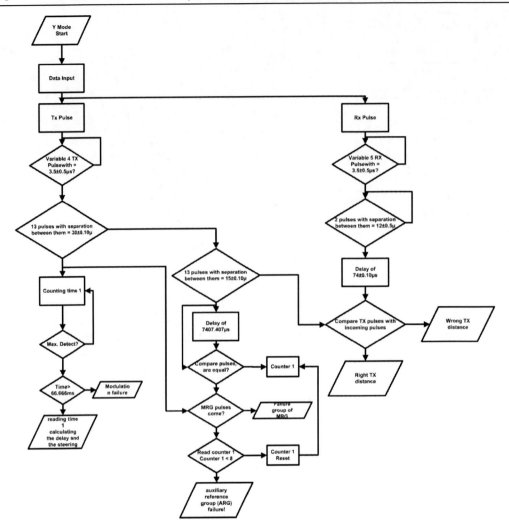

Figure 15. Knowledge model - Y mode

```
a0=Data(n,4); % Pulse level
n=n+1;
a1=Data(n,4);
if a0<2 & a1>3 t0=Data(n,1); % Detects rising edge
else
if a0>2 & a1<3 t1=Data(n,1); % Detects the falling edge
ta=t1-t0; % time between rising edge and falling edge
m=m+1;
V(m,:)=ta; % Column type vector with time of the pulse widths
endif
endif
endfor
save ("-text","Vector.csv","V") % saves the column vector V
"File saved"
```

5.1. Results analysis

The results of the performing of the statistical treatment are shown in table 3.

File name	Number of pulses	Mean	Variance	Standard deviation	Mode
Vector.csv	3346	4.19E-6	1.37E-12	4.57E-9	4E-6

Table 3. Results of the test

6. Conclusion

The knowledge model presented provides a method for understanding the TACAN system, defining rules to ensure that the system operates according to the standard. The rules help us to understand the operation and to establish guidelines for diagnosing faults, and this also contributes to a faster repair and restoration of the operation of the system.

This method is excellent for understanding the operation of the TACAN, because each part of it is analyzed independently of the others. Therefore, it provides a powerful support for both application areas: the maintenance and the education.

References

[Bin and Wei-kang, 2009] Bin, Z. and Wei-kang, W. (2009). Research on accurate measurement method of tacan azimuth based on curve fitting. In *5th International Conference on Wireless Communications, Networking and Mobile Computing, 2009*, pages 1–4.

[Chang et al., 2012] Chang, N., Prapinpongsanone, N., and Ernest, A. (2012). Optimal sensor deployment in a large-scale complex drinking water network: Comparisons between a rule-based decision support system and optimization models. *Computers & amp; Chemical Engineering*, 43:191–199.

[Chen et al., 2012] Chen, Y., Poon, S., Yang, J., Xu, D., Zhang, D., and Acomb, S. (2012). Belief rule-based system for portfolio optimisation with nonlinear cash-flows and constraints. *European Journal of Operational Research*, 223(3):775–784.

[Cheng et al., 1993] Cheng, W., Mar, A., Bowers, J., Huang, R., and Su, C. (1993). High speed 1.3 μm ingaasp fabry-perot lasers for digital and analog applications. *IEEE Journal of Quantum Electronics*, 29(6):1660–1667.

[Cimino et al., 2012] Cimino, M., Lazzerini, B., Marcelloni, F., and Ciaramella, A. (2012). An adaptive rule-based approach for managing situation-awareness. *Expert Systems with Applications*, 39(12):10796–10811.

[Civil, 2012] Civil, A. (2012). *Historia del tráfico aéreo*. Last checked: 2012-11-09.

[Cline et al., 2010] Cline, B., Brewster, C., and Fell, R. (2010). A rule-based system for automatically evaluating student concept maps. *Expert Systems with Applications*, 37(3):2282–2291.

[Cordón, 2011] Cordón, O. (2011). A historical review of evolutionary learning methods for mamdani-type fuzzy rule-based systems: Designing interpretable genetic fuzzy systems. *International Journal of Approximate Reasoning*, 52(6):894–913.

[Costa et al., 2012] Costa, C., Luciano, M., Lima, C., and Young, R. (2012). Assessment of a product range model concept to support design reuse using rule based systems and case based reasoning. *Advanced Engineering Informatics*, 26(2):292–305.

[deFaymoreau, 1956] deFaymoreau, E. (1956). Experimental determination of tacan bearing and distance accuracy. *Aeronautical and Navigational Electronics, IRE Transactions on*, ANE-3(1):33–36.

[Digital, 2012] Digital, A. (2012). *Historia de la aviación*. Last checked: 2012-11-11.

[ElAlami, 2011] ElAlami, M. (2011). Unsupervised image retrieval framework based on rule base system. *Expert Systems with Applications*, 38(4):3539–3549.

[F. Hayes-Roth, 1983a] F. Hayes-Roth, e. a. (1983a). *Building expert systems. Reading.* Addison-Wesley.

[F. Hayes-Roth, 1983b] F. Hayes-Roth, e. a. (1983b). Rule-based systems. commun. *ACM*, 28(9):921–932.

[Gao et al., 2011] Gao, J., Fu, C., Liu, Y., and Wei, Y. (2011). Behavioral modeling and emc analysis for tacan system. In *Microwave, Antenna, Propagation, and EMC Technologies for Wireless Communications (MAPE)*, pages 576–579. IEEE 4th International Symposium on.

[Hadeii et al., 2011] Hadeii, M., Ghorbani, A., and Khorramabad, S. (2011). Design and simulation of beam forming network for tacan radar. In *IEEE International RF and Microwave Conference (RFM)*, pages 49–53.

[Ishibuchi et al., 1999] Ishibuchi, H., Nakashima, T., and Morisawa, T. (1999). Voting in fuzzy rule-based systems for pattern classification problems. *Fuzzy Sets and Systems*, 103(2):223–238.

[K. Khalili-Damghani, 2011] K. Khalili-Damghani, S. Sadi-Nezhad, M. A. (2011). A modular decision support system for optimum investment selection in presence of uncertainty: Combination of fuzzy mathematical programming and fuzzy rule based system. *Expert Systems with Applications*, 38(1):824–834.

[Kayton and Fried, 1997] Kayton, M. and Fried, W. R. (1997). *Avionics navigation systems*. John Wiley, second edition edition.

[Sandretto, 1956] Sandretto, P. C. (1956). Development of tacan at federal telecommunication laboratories. *Electrical Communication*.

[Sanz et al., 2011] Sanz, J., Fernández, A., Bustince, H., and Herrera, F. (2011). A genetic tuning to improve the performance of fuzzy rule-based classification systems with interval-valued fuzzy sets: Degree of ignorance and lateral position. *International Journal of Approximate Reasoning*, 52(6):751–766.

[Shi et al., 1996a] Shi, Y., DJ.Olson, Bechtel, J., Kalluri, S., Steier, W., Wang, W., Chen, D., and Fetterman, H. (1996a). Photoinduced molecular alignment relaxation in poled electro?optic polymer thin films. *Applied Physics Letters*, 68(8):1040–1042.

[Shi et al., 1996b] Shi, Y., Wang, W., Bechtel, J., Chen, A., Garner, S., Kalluri, S., Steier, W., Chen, D., Fetterman, H., Dalton, L., and Yu, L. (1996b). Fabrication and characterization of high-speed polyurethane-disperse red 19 integrated electrooptic modulators for analog system applications. *IEEE Journal of elected Topics in Quantum Electronics*, 2(2):289–299.

[Shi et al., 1997] Shi, Y., Wang, W., Lin, W., Olson, D., and Bechtel, J. (1997). Double-end crosslinked electro-optic polymer modulators with high optical power handling capability. *Applied Physics Letters*, 70(11):1342–1344.

[Wikipedia, 2011] Wikipedia (2011). *Sistema VOR*. Last checked: 2011-02-27.

[Wikipedia, 2012a] Wikipedia (2012a). *Sistema TACAN*. Last checked: 2012-11-10.

[Wikipedia, 2012b] Wikipedia (2012b). *Tactical air navigation system*. Last checked: 2012-11-10.

[Wood-Hi, 1994] Wood-Hi, C. (1994). Effect of carrier lifetime on mode partition noise in multimode semiconductor lasers. *Photonics Technology Letters, IEEE*, 6(3):355–358.

SUBSECTION 2: SAFETY AND HEALTH RISK

In: Ships and Shipbuilding
Editor: José A. Orosa

ISBN: 978-1-62618-787-0
© 2013 Nova Science Publishers, Inc.

Chapter 5

RESEARCH ABOUT THE RISK OF EXPLOSION ON BOARD FROM USING LIQUEFIED NATURAL GAS AS FUEL

Saturnino Galán, José A. Orosa, Angel Rodríguez and José A. Pérez*

Department of Energy and Marine Propulsion. University of A Coruña, Department of Industrial Engineering II. University of A Coruña, Spain

ABSTRACT

The shipping industry is currently under pressure to comply with international regulations that are becoming more and more strict with environmental matters, to which must be added the increase in fuel costs, so that the shipping sector changes the trend in fuel consumption, supported by IMO, to reduce the environmental impact caused by merchant ships. LNG used as fuel, is the means by which to reduce emissions from ships, because every voyage, compared with those using conventional fuels, reduces the environmental impact of 100% of SO_2 emissions, 90% of NO_x emissions and almost 30% of CO_2 emissions, which translates to a significant reduction in emissions, which means that this is the optimum new fuel for use on ships. If we pay attention to the components of a propulsion system for fuel gas, it becomes imperative for us to pay special attention to the security aspects of LNG. In this chapter, we have studied the effects of accidental explosions on board ships, and to take security measures required to ensure adequate structural integrity of the ship in case of an accident.

1. INTRODUCTION

The European Parliament adopted the European Directives 1999/32/EC, 2005/33/EC, and 2010/769/UE relating to reduction of sulphur emissions from ships, with the aim to improving air quality in coastal areas. The International Maritime Organization (IMO) [1],

---

* Corresponding Author address: E.T.S.NyM. Universidade da Coruña. Email: jaorosa@udc.es.

through the revision of MARPOL, Annex VI, calls for stricter limits by way of particularly stringent sulphur controls when using marine fuels in areas of SOx emission control in sulphur emission control areas (SECA), and these limits (1.00% as of July 1, 2010 and 0.1% as of January 1, 2015) and in sea areas outside the SECA (3.50% as of 1 January 2012 and, in principle, 0.50% as of January 1, 2020) [1]. As a result of this, the shipping industry is currently under pressure to comply with international regulations that are becoming more and more strict regarding environmental matters. Added to this is the increase in fuel costs, which has resulted in the shipping sector changing the trend in type of fuel ussed, which move is supported by IMO, to reduce the environmental impact of pollution by merchant ships.

The shipping sector, in search of suitable alternative fuels, has found liquefied natural gas (LNG) to be the ideal replacement for traditional fuels.

LNG, used as fuel, reduces emissions from ships because every voyage of such ships, when compared with ships using conventional fuels, reduces the environmental impact thus: 100% SO_2 emissions; 90% NOx emissions; and almost 30% CO_2 emissions, which translates to a significant reduction in emissions and pollution, thus proving that this is the optimum new type of fuel that should be used on ships [2].

Natural gas generates no waste or by-product, and therefore, production and release of particles into the atmosphere is also non-existent. In addition to the environmental benefits, the use of gas as fuel has the potential to reduce fuel consumption, which relates directly to the economics and cost of transportation operations in the shipping industry.

If we are to consider the various components of a propulsion system in the use of LNG as fuel, it is imperative for us to pay special attention to the storage systems and security aspects of LNG [2].

2. STANDARD SYSTEM OF A SHIP USING LNG AS FUEL

Both, new vessels and those under construction, suffer from a drawback in their propulsion systems and fuel. The standard system on a ship should consist of the following:

- Installing fuel tanks.
- Gas processing equipment.
- Security systems.
- Engines, suitable or converted.
- Ventilation outlets in safe locations, and separating the output of the exhaust gases.

Unlike the traditional storage tank systems, fuel storage vessels and the gas industry mainly used storage containers, like bottles or containers. One of the goals should be to ensure a safe and secure storage that allows normal operations, without compromising either the operation of the ship or the safety of its crew and passengers [3].

According to one of the world's major engine manufacturers, Wärtsilä, for an average consumption during about 14 days voyage, and at the rate of 50 m^3 per day, keeping a safety margin of 10%, the storage capacity of the LNG on board a ship must be roughly 770 m^3 storage system of gas (LNG). This requires appropriate conditions to ensure the safety of the facilities and the structural integrity of the ship in case of an accident [4].

3. LOCATION OF DIFFERENT LNG STORAGE SYSTEMS ON BOARD

Vessels, as they begin to switch over to LNG as fuel, must possess the necessary infrastructure and measures regarding structural integrity and resistance in the hull, or in the storage space intended for LNG.

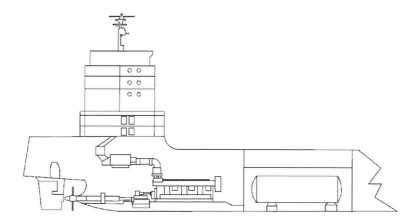

Figure 1. A vessel with LNG as fuel.

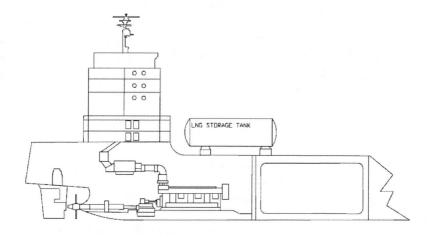

Figure 2. LNG storage tank.

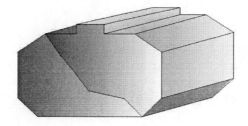

Figure 3. LNG storage tank.

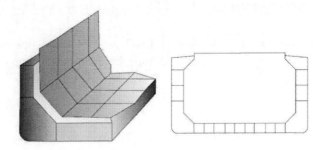

Figure 4. Prismatic LNG tank as used in LNG carriers.

The LNG storage system on board a vessel can be planned in various ways, either in cylindrical tanks, which can be of the insulated type, with a storage pressure of between 5 and 10 bar, or prismatic tanks, membrane type, such as those used in transporting LNG in tankers.

Storage tanks may also be of membrane type, such as those used in transporting LNG in tankers, or prismatic tanks, as shown in Figures 3 and 4.

The location of the LNG storage tanks on ships, such as container ships, passenger ships and ferries, is one of the key issues in safety and integrity in the event of any accident, especially in ferries, where there is almost no alternative to placing them under cover.

4. Dangers of Using LNG as Fuel

Gas, as fuel, has certain risks and dangers that need to be addressed, since the operation of the vessels and facilities on board require flexibility and new approaches to solving new problems and needs, addressing new challenges, but without neglecting or compromising the safety of goods, facilities, human beings and the environment.

The potential hazards should be assessed and taken into account when designing (or redesigning) a vessel using LNG as fuel. These are:

- Gas tankers are not designed for this purpose.
- Location of the fuel storage tanks.
- Treatment of gases and vapours from LNG.
- Fuel supplies.
- Fire or explosion from ignition of gas leaks in the LNG bunkering system.
- Gas engines (including pipelines).
- Spray.
- Exhaust system.
- Ventilation.
- Fire/explosion in other areas.
- LNG leakage causing weakening or loss of structural integrity/brittle fracture of structures.
- Asphyxiation due to lack of oxygen in the event of a spillage of LNG in an enclosed or semi-enclosed area.
- Human contact with surfaces or substances at extremely low temperatures.

- Collisions can damage the containment system of LNG/gas and, in extreme cases, also the LNG storage tank itself.

New ships using LNG as fuel must have suitable protective measures necessary for safeguarding the port/town, or in the room, or in the section of the vessel for housing the storage or containment of LNG, in the event of an explosion, or failure of one of the fuel containers [5].

5. FUEL SUPPLY NETWORKS (BUNKERING)

Grids (bunkering) vessels have not been developed yet, but they should complement the security measures. Also, these will involve a review of all safety and operational procedures, since the transfer of a cryogenic fluid that is not only combustible, but highly volatile, requiring special operations in the case of LNG tankers, is in the charge of officers and engineers in the loading and unloading operations of these ships [6].

6. SOURCES OF KNOWLEDGE

Since the 1960s, tankers have provided valuable knowledge to us with regard to explosions in the industry, and we now have a wealth of knowledge that gives us a good idea of the dangers we face in the event of an explosion of an LNG container.

We need to determine the risk of exploding shells, and to assess appropriate measures taken.

Like the industry in general, in the LNG storage systems on ships, among others, there may be situations of pressure, gas containment, explosion, sudden failure of material, deposits, and so on, so that it is necessary to undertake analyses of these situations, and arrive at estimates of the impulse generated over pressure, and in the event of explosion of the gas containers. We can use the methods of Multi-Energy and Baker Strehlow.

LNG is transported at $-165°C$, thus, any human physical contact with the metal container can be extremely dangerous; while stainless steel will remain ductile, carbon steel and low-alloy steels will become fragile and even brittle, so that fractures in the material are likely to happen if exposed to such extremely low temperatures.

In the case of irrigation collision is considered feasible to construct, locate and protect fuel tanks on ships powered by LNG.

For accidents occurring in the industry, we provide a source of information, since most of the substances and liquids are handled, stored and processed in siphoning tanks and containers.

Accidental explosions have occurred, and continue to occur, during process operations in chemical and petrochemical plants, when cleaning the fuel tanks of tankers, storage products, and generally in any field associated with combustible gasses. These have always been, and will remain, potential risks of accidents [7].

7. RISKS AND DANGERS

The existence of a risk does not necessarily mean that it can lead to some harm, if not the possibility of damage. Different kinds of damages that can occur can be physical, chemical, or because of unforeseen circumstances. The reasons may be the origin or source of the explosions, and can be many and varied. But, the most common, it should be noted, are:

- The change of a property or a dimension of the processed material or object.
- The failure of one (or more) parts of some appliances, security systems and components.
- External disturbances, such as: shock vibrations, electromagnetic fields, etc.
- Errors or design deficiencies.
- Disruption of power supply or other services.
- Loss of control by the operator (especially for portable and mobile controls).
- Factors related to security.
- Factors related to the environment and working conditions.

The decision, therefore, may occur as a result of human error, faulty materials, appearance of abnormal operating conditions or deviation from normal operation. The most common consequences are usually:

- Rupture of pressure equipment.
- Explosion of equipment.
- Leaks and spills of LNG.
- Fires.

Reducing risks in handling and dealing with this type of equipment or facilities is virtually impossible, so we try to reduce risks to the least possible extent, limiting their consequences in order to protect people, property and the environment. These risk situations sometimes materialise in accidents that could result in devastating consequences.

8. RECENT ACCIDENTS IN INDUSTRIAL INSTALLATIONS

Recently, there have been a series of accidents at different locations around the world in petrochemical plants, and the examples given below show the devastating effects of such explosions. These explosions involved combustible chemical and petroproducts, so in addition to the effects of the pressure wave caused by the explosion must be added the effects of fire, and the eye-witness descriptions give us a fairly good idea of how devastating such a calamity can be.

On August 14, 2012, an explosion occurred in the refinery of the industrial city in south Madero Tamaulipas State, Mexico, in the refinery belonging to the company, Pemex. No casualties were reported, only property damage. The oil company reported that the incident occurred in a boiler plant hydrodesulphuriser U 300 in the refinery. In addition to this, the explosion caused a fire at the gasoline heater BA 302 Unit 300 (hydrodesulfurisation).

There were several causes for the explosion, including the action of lightning. The oil company, Pemex, suggests that the probable cause of the explosion could have been caused by an overpressure inside the heater BA hydrodesulphuriser plant 301 U 300. It also notes that because of this the heated BA 302 was damaged in the convection section, causing damage to the pipe rack. The explosions were heard at a radius of 70 kms. Government sources, in a first estimate, calculated that the refining plant, before the accident, could produce 50,000 barrels of gasoline per day, but was unable to determine how many days it would now be out of the refinery operation [8–14].

On September 18, an explosion and subsequent fire occurred in the gas plant of Pemex Exploration and Production, 19 kms from Reynosa in Tamaulipas State. In this installation, concentrated gas and condensate fields are produced by the Burgos Basin, in order to separate and measure them. Once separated, the gas is delivered to the Burgos Gas Processing Complex where it undergoes a process called sweetening. This tragic accident took the lives of 30 people. The damage was substantial. An idea of the range of the explosion is shown in Figure 5 [15–16].

On August 25, 2012, a gas leak caused an explosion at the refinery, Amuay Paraguana Refining Center, in Falcon State, Venezuela. This tragic accident took the lives of 42 people. The magnitude of the explosion was such that initially, it was compared with an earthquake or a natural disaster. More than 200 homes near the complex had been affected by the explosion.

Government sources of the Ministry of Energy and Petroleum of Venezuela, have said that the accident stemmed from a leak of olefins (gas used in the manufacturing process of gasoline). Because of weather conditions, this leak created a gas cloud which did not disperse, but formed a cloud of gas which then ignited. Other sources suggest that the lack of maintenance, lack of investment and violation of safety standards, may have been the main causes for the accident. The explosion caused fires in at least two tanks of the refinery and the surrounding atmosphere. Some idea of the range of the explosion is shown in Figure 6 [17–20]. On August 6, 2012, in Richmond, California, a fire originated in the Chevron refinery in the town. The proximity of the population forced the authorities to decide to order everyone to stay inside their homes with their windows closed. The fire started in the unit's oil refinery No. 4 (hours before an inspection team had detected a leak of diesel on a line drive). There were no fatalities, but one worker was seriously injured. An idea of the force of the explosion can be had from Figure 7 [21–22]. On 1 October 2012, there was an explosion of a fuel tank in the city of Winnipeg, Canada. This explosion was the result of a massive fire in the premises of the company, International Speedway Ltd., a company that sells methanol as fuel for racing cars. As a result of the fire, 100 nearby homes were evacuated, many of them 1 km away from the fire. Moments before the methanol tank explosion, firefighters were advised to take cover behind something solid, like dumps and rail cars. Sources confirm that the company intended to carry out fuel storage in an unauthorised manner in rail cars on an adjacent siding, and that their activity did not have the requisite permission from the authorities.

The rail cars with 100,000 litres of fuel stored, could have been affected by the fire with consequences that would have been devastating. Estimates of the damage put the total cost of the incident at about 15 million dollars. The owners say they had passed all safety inspections, and had complied with regulations. Fortunately there were no casualties [23–24].

On October 6, 2012, a truck carrying liquefied petroleum gas overturned on a highway in Yunaling, Hunan Province in China, and caused a huge explosion.

Figure 5. Recent accident in Tamaulipas, Mexico.

Figure 6. Recent accident in Reynosa, Mexico.

Figure 7. Recent accident in Amuay, Venezuela.

Figure 8. Recent accident in the Chevron plant, Richmond, California, USA.

Figure 9. Recent accident in Winnipeg, Canada.

Figure 10. Recent accident in Yunaling, China.

Figure 11. Recent accident in La Coruña, Spain.

Tragically, two people died in the accident, but it was the separation of the fuel tank container that caused the explosion that killed three firefighters who tried to help in the accident. According to local media, the blast was so powerful that it uprooted trees and destroyed several vehicles and fire trucks. An idea of the range of the explosion is shown in Figure 10 [25]. On October 10, 2012, a fire occurred in the FCC plant in Repsol Industrial Complex in La Coruna, Spain. The causes of the fire. An injury was. An idea of the range of the explosion is shown in Figure 11 [26–28].

9. DEVASTATING POWER OF EXPLOSIONS

These examples of recent explosions serve to give us an idea of the devastating power of explosions and the damages they can cause, even with regard to human lives. People unconsciously and unwittingly accept certain risks without examining the details and safety aspects too much, either because of ignorance, because they are little-known risks, or due to sheer carelessness and blind acceptance, as these are situations or processes being fully utilised. In both cases, one has to assume a certain risk, but this should not be voluntary, or at least should be semi-controlled, and economic values or other costs that reduce the extent of adoption of safety measures should never be allowed to prevail.

We have seen that as a rule, an explosion is followed by a trail of damage that can be:

- Structural, caused by shock waves and the impact of fragments.
- Injuries or deaths caused by the explosion and impact of debris.
- When flammable substances come into play, thermal radiation and its effects.

10. ANALYSIS OF THE EFFECTS OF EXPLOSIONS

To analyse the effects of all these explosions, a brief study which evaluates and classifies risks is required for a description of the phenomena that come into play in the blasts. Evaluating the explosions and their effects is a study of the specific characteristics of the blasts, which are pressure waves, and the effects thereof on people and structures. It also sets out the various calculations that are used to analyse the consequences, and to establish the most appropriate preventive measures. The method can be applied to all those pressure vessels, facilities and equipment used in industry, in ships and shore units, where there is a risk of an explosion occurring. By the laws, rules, codes, regulations, rules of classification societies, which are mandatory, not about limiting the use and the use of equipment and facilities, if they are limited only to establish basic rules of activities with the aim of regulating particular security on site. As in industry, so also on ships, we can find cases of avoidable accidental explosions. These can be caused by the facilities and equipment that contain:

- Flammable products.
- Gas mixtures.
- Combustible dust.
- Clouds of vapour.

As with all installations and pressure equipment where explosions may occur, containers of compressed gases, liquefied gases, liquids, combustible liquids, or liquids, were overheated [29].

10.1. Risk Assessment

Following the methodology established in the UNE EN ISO 14121-1, the risk assessment must be carried out for each particular case, including:

- Identification of hazards.
- Determining whether it will produce an explosive atmosphere, and the extent involved.

- Determining the presence and the possibility that ignition sources exist that are capable of igniting the explosive atmosphere.
- Determining the possible effects of an explosion.
- Estimating the risks.
- Considering measures to reduce the risks.

Damage can be caused by these dangerous phenomena:

- Mechanical phenomena: pressure waves and shells.
- Thermal-type phenomena: heat radiation.
- Phenomena of chemicals: toxic clouds [30].

10.2. Explosions

From the examples of various explosions described above, each image is immediately associated with destruction. An explosion is a phenomenon associated with a sudden release of energy to the atmosphere (or in water or below ground) which produces a pressure wave that moves radially away from the central source, while it loses energy over distance and time. The energy release is quite rapid and concentrated, for the wave generated is audible.

Explosions are defined by the rate of reaction. The propagation speeds are in excess of normal (1 m/s). Pressure waves generated are high to low level, depending on whether they are developed in open or closed conditions that will help in its dissipation, or suffer an even greater process compression. The generation of high pressures typical of explosions, are characterised by the value of the maximum lift speed of the pressure, obtained by the angle between the tangents of these elevations, and that constitutes a comparison parameter, which can be represented in curves that distinguish one type of explosion from another [31].

As seen in the above examples, ignition of certain fuels, and their subsequent combustion, dramatically manifests, resulting in intensity and destructive effects, and these cases are referred to as explosions [32].

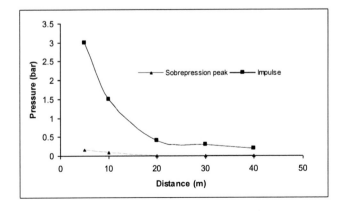

Figure 12. Recent accident in La Coruña, Spain.

10.3. Origin of Bursts

- Explosions that result from a fire.

Such explosions are not to be associated with fire, but only if they are caused by it. This occurs when the flames impinge on the outside of a vessel or piping, thus heating it. The heat generated by a fire of any kind can cause explosions.

The partial opening of the system itself may be due to material failure of the container by heat (especially the area which is in contact with the vapour phase not cooled by the boiling liquid). This is part of the heating leading to increases in pressure and temperature, which also contributes to the explosion.

- Explosions that result from other explosions.

An explosion can trigger a leak, fire and other explosions. On the one hand, the blast can distort and even destroy other equipment or containers. Furthermore, the projectiles can come from an explosion, and can cause similar effects. It is not necessary to generate damage to consider this phenomenon as an explosion. The energy released initially may have been stored under a variety of forms: electric, nuclear, chemical or pressure.

10.4. Types of Explosions

The BLEVE is defined as a failure by corrosion of a boiler. Breakage of a container is defined as a burst pressure vessel or mechanical failure of a container containing pressurised gas.

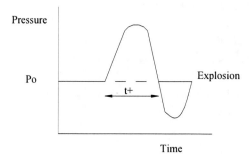

Figure 13. Explosion deflagrating.

A pressure vessel is one that contains pressurised gas. Relief system failure (safety valve) happens during overpressure in a pressure vessel. Fast transmission phase is the rapid transition phase of hot oil in a bowl of water, LNG and water mixture.

10.4.1. Explosions Deflagrating

Such explosions are those which starting from reaction rates between 1 m/s and the speed of sound, create pressure waves that do not reach values higher than 10 kg/cm^2 pressure.

This type of explosion is the result of a dusty atmosphere or combustible liquid vapours. An explosion deflagrating will have less impact if it occurs in open areas that allow the dissipation of pressure waves that are generated, as opposed to an explosion that occurs in a confined space.

11. EXPLODING CONTAINERS

The causes that can lead to fracture of a container are diverse. However, there are basically two reasons for the failure of a pressure vessel, which are:

- For structural weakness (corrosion, erosion, fatigue, defective materials, etc.), by external shocks.
- For increased pressure due to different causes (overfill, runaway reaction, internal explosion, failure of a safety system or control, or an internal explosion).

A third reason is that of a combination of the above two phenomena. Thus, the external heating by fire produces an increased pressure inside the container, and the weakening and failure of the materials which constitute it. These explosions are much more violent and can cause great harm, especially when the contents are flammable, and the pressure wave which is generated at the outbreak of the container must be added to the ignition and, consequently, the fire from the product inside the container.

11.1. Explosions from Physical Causes

11.1.1. Expansion of a Gas under Pressure
When a container containing a pressurised gas bursts, the only energy source available for fragmentation and generation of the pressure wave is the expanding gas. The energy released will depend, therefore, on the storage conditions, primarily pressure and quantity, of the substance involved.

It should be clarified that explosions have been defined variously, as the sudden variation of pressure is called overpressure.

11.2. BLEVE

The BLEVE, or 'liquid expanding vapour explosion boiling', is a special kind of explosion, very characteristic and devastating, that occurs in tanks or containers in which fluid is stored under pressure. Contrary to what might be thought initially, these explosions do not always have associated thermal effects. This depends on the type of substance involved, for example, with flammable liquids.

11.2.1. Root causes of BLEVE

The three main reasons for these being produced are:

- The liquid is superheated.
- There is a sudden and rapid drop in pressure (sudden depressurisation), which can be caused by cracks (bursting of container, mechanical impacts on the container), or ruptured discs action, directly on the reservoir.
- Related with the previous two, that the pressure and temperature specific to each product allow spontaneous nucleation (and simultaneous sudden formation of bubbles throughout the mass of the liquid), causing very rapid vaporisation.

11.2.2. Fast Transmission Phase

The extraordinary increase in volume that causes a liquid to vaporise, which is about 1,700 times in the case of water, and about 250 times in the case of propane over existing steam expansion, will cause a pressure wave (explosion/blast), and the breaking of the container into several pieces which can be launched up to a considerable distance.

In such a situation, the overpressure caused by the rapid transmission phase caused by the interaction of LNG and water are also typical in installations which operate with LNG regassification, such as liquefaction and transportation facilities in LNG ships.

Very small quantities of liquid can turn into large volumes of gas. One volume/unit of LNG can produce approximately 600 volumes/units of gas. Approximately 1 m^3 of liquid LNG produces approximately an order of 600 m^3 of gas.

Natural gas, like all other gaseous hydrocarbons, is flammable. At ambient conditions, this happens when the mixture of air and flammable gas content is between about 5% and 15% by volume.

12. EFFECTS OF THE PRESSURE WAVE

In the case of explosions, the factors that come into play, and that need to be analysed in order to estimate the consequences are:

- Training the flames.
- Thermal radiation.
- Energy released in the explosion.
- Overpressure wave.
- Projected fragments.
- Container breakage.
- Projection of expansive steam, which can drag liquid particles in mist form.
- Hazardous substance emissions.

Knowing the impact assessment and the scope thereof, it can:

- Establish safety distances.
- Know and set protection modes.

- Set resistance of the air envelope and its supporting structures.
- Take decisions to make changes in the system or process to mitigate the risks.

12.1. Response of Structures to the Pressure Wave

At the moment of explosion, the pressure wave generated, as well as the basic characteristics of pressure and impulse, interact with the surroundings, displacing and/or compressing whatever is in their path, so that depending on their position in the explosive charge, the size, shape, nature and composition generate huge forces and results. These generate a 'domino effect' of damage and devastation on equipment, installations and structures, and often humans.

Damage caused by pressure waves on structures depends on the pressure and pulse that form projectiles. Therefore, the two major considerations in assessing the effects of a shock wave on a structure are:

- Predict increased pressure of an explosion.
- Estimate the reaction of the structures, and the damage that occurs within them.

This is often used with historical data tables obtained. It is one of the most used methods to have an idea of the magnitude of the consequences of an explosion. By way of example, the following table presents damages relating to certain overpressures.

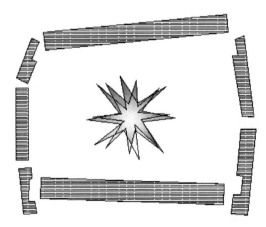

Figure 14. Damage caused by pressure waves.

Table 1. Structural damages and its related overpressure

Structural Damage	Overpressure
Total destruction	from 1 to 2 bar
Destruction of heavy machinery >3500 kg	0.68 to 0.70 bar
Dump rail cars	0.43 bar
Minor damage to heavy machinery >1500 kg	0.27 bar
Steel structures torn or displaced	0.27 bar

19. Methods of Calculating and Predicting the Extent of Explosions

To arrive at an estimate of the effects of an explosion, it is necessary to have some comparative index; thus, there are several applicable methods depending on the substances that come into play, the type of environment in which the explosion occurs, and the conditions surrounding the environment. There are several methods for calculating, or models to describe the effects, of an explosion. Depending on the type of explosion and the products involved, these methods of calculation are set out in the following table:

As can be seen, there are different types of calculations, depending on the types of substances and conditions. The three most widely used computational models to study and describe the effects of an explosion are: method of equivalent TNT, multi-energy method, and the Baker method.

By way of example, the effects of the explosion of a pressure vessel of compressed air from a volume of 12 litres and an inner storage pressure of 200 bar, similar to a breathing apparatus, calculated with the method of equivalent TNT (the least accurate of all for calculating an explosion in a container), are shown in the figure and table appearing below [33].

Table 2. Calculation methods of different types of explosions

Type of explosion	Calculation method
Detonation of explosive	TNT equivalent
Explosion of steam clouds	Method based on TNT
	Method based on the characteristics of the explosions
	Fluid dynamic method
	Multi-energy method
	Others
Bursting of containers	Fluid dynamic method
	TNT equivalent
	Baker
BLEVE	TNT
Toxic or flammable cloud dispersion	

Table 3. Calculation methods of different types of explosions (II)

Real distance (m)	Distance for self-contained breathing apparatus (SCBA)	Sobrepression peak (bar)	Impulse (scaling)	Impulse (bar.s)	Duration (scaling)	Duration of the positive phase (s)
5	10.863	0.150	6.5	2.99	3.5	1.610
10	21.727	0.060	3.3	1.51	4.5	2.071
20	43.454	0.032	0.9	0.41	6.3	2.899
30	65.181	0.022	0.7	0.32	6.5	2.991
40	86.909	0.020	0.6	0.27	6.8	3.129

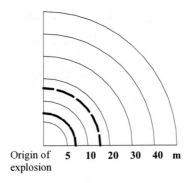

Figure 15. Effect of an explosion.

Table 4. Effect of an explosion

Distance (m)	Distance for self-contained breathing apparatus (SCBA)	Sobrepressure ΔP (bar)
2	0.767	0.7000
3	1.151	0.3000
4	1.534	0.1800
5	1.918	0.1500
10	3.836	0.0500
20	7.673	0.0250
30	11.510	0.0150
40	15.347	0.0100
50	19.184	0.0080
60	23.020	0.0065
70	26.857	0.0060
100	38.368	0.0030

At a distance of 5 m, 0.15 bar overpressure, and a boost of 2.99 bar • s; this implies that, in this case, the damage at that distance might be repairable. However, if the calculations are made with the Baker method, these differences show:

The overpressure generated at a distance of 5 m is 0.33 bar, which involves the partial destruction of structures. Finally, these results show how the failure of a small pressure vessel can cause partial destruction of larger structures [34].

CONCLUSION

No doubt, future maritime industry and transport of goods by sea, will see the passing and adoption of LNG as fuel, but this should necessarily involve a substantial increase in security measures, because the inherent characteristics of LNG involve some real hazards.

To minimise these risks, all necessary measures to prevent fire or explosion in areas, like the engine room, should be as simple as possible, and non-essential items removed from the spaces around the main engine area; and the installation of fire extinguishers, wherever necessary, to ensure the safety of people and the facilities.

The designers, builders, owners and classification societies should set standards to ensure the safety of the facilities. We have a source of knowledge and technology, extensively tested and proven, for use of LNG as the fuel system in existing tankers. They must adopt a system, like in gas carriers, to operate in other ships, such as container ships, bulk carriers, and even large passenger ships.

In accidents in the oil industry, we have highlighted the devastating effects of explosions, so by applying forecasting systems and appropriate calculations, it should be possible for us to take necessary steps to minimise the effects of explosions on board merchant ships. For shipping companies, the loss of cargo space can be an important factor in the case of LNG storage facilities for fuel on board the vessel, but we should never allow any dilution in safety and security measures, whether human or material.

REFERENCES

[1] International Maritime Organization (2009). Revised Marpol Annex VI: Regulations for the Prevention of Air Pollution from Ships and Nox Techical Code 2008.

[2] DNV, Marpol 73/78 Annex VI Regulations for the prevention of air Pollution from Ships. Tecnical and Opertation Implications. www.dnv.com/binaries.

[3] Germanischer Lloyds. (2012). Nonstop the magazine for customers and businees partners. *ISSUE,* 02-2012.

[4] Oskar Levander. Wärtsila. Turning the page in ship propulsion, by switching to LNG. Gas as fuel for propulsion of ships- status and perspectives. Copenhagen, March 3rd, 2008. www.wartsila.com.

[5] Germanischer Lloyds. Costs and benefits of LNG as Ship fuel for container vessels. www.gl-group.com (Accessed December 2012).

[6] Germanischer Lloyds. LNG powering the future of shipping. www.gl-group.com (Accessed December 2012).

[7] Lloyd´s Register. (2012). Gas Technology. A special report on marine and offshore gas solutions.

[8] Lloyd´s Register. LNG fueled deep sea shipping. The outlook for LNG bunker and LNG fuelled newbuild demand up to 2025. August 2012 www.lr.org.

[9] Sinembargo. http://www.sinembargo.mx/14-08-2012/332878 (Accessed December 2012).

[10] Unonoticias. http://www.unonoticias.com (Accessed December 2012).

[11] El universal http://www.eluniversal.com.mx/notas/864220.html (Accessed December 2012).

[12] Noticias terra. http://noticias.terra.com.mx/desastres/pemex-controla-incendio-en-refineria-de-tamaulipas,3233e982b1529310VgnVCM5000009ccceb0aRCRD.html. (Accessed December 2012).

[13] Milenio. http://www.milenio.com/cdb/doc/noticias2011 (Accessed December 2012).

[14] Negociosreforma. www.negociosreforma.com (Accessed December 2012).

[15] http://www.minci.gob.ve/2012/09/18/mexico-registran-explosion-y-un-incendio-en-refineria-de-pemex. (Accessed December 2012).

[16] La republica. http://www.larepublica.es/2012/09/explosion-en-refineria-de-pemex-mexico-deja-10-muertos-y-40-heridos (Accessed December 2012).

[17] Noticialdia. www.noticialdia.com (Accessed December 2012).

[18] www.ft.ci.org (Accessed December 2012).

[19] Jfblueplanet. http://jfblueplanet.blogspot.com.es/2012/08/amuay-la-explosion-mas-grave-en-el.html. (Accessed December 2012).

[20] Noticiascentro. http://www.noticiascentro.com/2012/explosion-de-madrugada-de-amuay-destruyo-la-mitad-de-la-principal-refineria-nacional. (Accessed December 2012).

[21] Sfgate. http://www.sfgate.com/bayarea/article/Fire-at-Chevron-refinery-in-Richmond-3767221.php#photo-3293688. (Accessed December 2012).

[22] Explosion At Chevron Refinery In Richmond. www.chevron.com News (Accessed December 2012).

[23] http://www.winnipegfreepress.com/local/speedway-compliance-an-issue-after-fuel-fire-172417911.html. (Accessed December 2012).

[24] http://www.thestar.com/news/canada/article/1266575--winnipeg-fuel-plant-issues-statement-about-explosion-says-safety-paramount (Accessed December 2012).

[25] Dashcamaccidents. http://www.dashcamaccidents.com/chinese-gas-tanker-leaks-causes-explosion. (Accessed December 2012).

[26] El país. http://ccaa.elpais.com/ccaa/2012/10/10/galicia/1349888243_088986.html (Accessed December 2012).

[27] La opinion. http://www.laopinioncoruna.es/coruna/2012/10/11/incendio-pone-alerta-refineria-repsol/654004.html. (Accessed December 2012).

[28] http://www.lavozdegalicia.es/noticia/coruna/2012/10/10/aparatoso-incendio-refineria-coruna/0003134988602039489021.htm (Accessed December 2012).

[29] J.Mª Storch de Gracia. T García Martín. (2008). Seguridad industrial en plantas químicas y energéticas. Fundamentos, evaluación de diseño y riesgo. Ediciones Díaz de Santos.

[30] UNE EN ISO 14121-1 Seguridad de las máquinas. Evaluación del riesgo

[31] Joaquim Casal. (1999). Análisis de riesgo en instalaciones industriales. Ediciones UPC.

[32] Saturnino Galán Fontenla. PFC FNB UPC Explosión de Equipos a Presión Análisis de Riesgos y Consecuencias. http://upcommons.upc.edu/pfc/handle/2099.1/15993.

[33] Methods for the determination of posible damage. TNO. (1989). Green Book.

In: Ships and Shipbuilding
Editor: José A. Orosa

ISBN: 978-1-62618-787-0
© 2013 Nova Science Publishers, Inc.

Chapter 6

RESEARCH ABOUT SAFETY OF WIRE ROPE ON BOARD

Saturnino Galán, José A. Orosa, Angel Rodríguez and José A. Pérez*

Department of Energy and Marine Propulsion
University of A Coruña, Spain
Department of Industrial Engineering II. University of A Coruña, Spain

ABSTRACT

Currently, the use of steel cables comprising the entire industry in general and its use, is extended to a large number of applications and uses.

The scope of analysis is extended to all those steel cables with magnetic properties commonly used in the wider industry, with particular interest in the maritime and port industry, as well as all cables used in the sectors that are encompassed within the following fields.

Today, because of the huge security controls and measures set out in the working world, the reliable and safe use of wire ropes is crucial for operations on land and at sea.

Therefore, the safety of the cables has to remain a constant priority for employers, wire rope operators and safety authorities. In accordance with this, in this chapter, the effects of the use of steel cables in ships is done with the aim to be guide for engineers and researchers.

INTRODUCTION: USING STEEL CABLES ON SHIPS AND OFFSHORE INDUSTRY

Steel wire ropes are used on seagoing vessels for any of a number of purposes. They are, for instance, fitted to cranes to hoist and move loads. They may be used on winches below deck, to lay pipes and for construction work in deepwater fields.

---

* E-mail: jaorosa@udc.es.

Steel wire ropes are used in the different tasks performed on board ships, including the following:

- Mooring lines in large vessels (e.g., tankers).
- Cables for heavy cargo-lifting cranes on board.
- Cables for lifting persons inside the vessel (e.g., elevators).
- Tow ropes and towing tugs.
- Hoisting cables in RoRo vessels.
- Cables for lifting cargo gateways in RoRo vessels.
- Cranes.
- Fishing and aquaculture.
- Exploration units in oil and gas rigs.

1. ANALYSIS OF A WIRE ROPE

In order to arrive at a correct analysis of the condition of a steel wire rope and take a correct decision, you need to have some previous knowledge of the following elements:

- Elements that constitute the cable.
- Wire rope classifications.
- Faults most commonly found in cables, and criteria for their discard.
- Techniques for inspection of steel cables.
- Magnetic inspection operating principle.

2. ELEMENTS THAT CONSTITUTE A WIRE ROPE

2.1. Wire Ropes

A steel wire rope consists of multiwire strands laid helically in one or more layers around a core. The way the wires are laid to form the strands, the way the strands are laid about the core, the core construction, and the materials and coatings used for the components contribute to the overall properties of the rope. We can also say that the steel cable is a simple machine that consists of a set of elements that transmit force, motion and energy between two points in a predetermined manner to achieve a desired end [1].

2.2. Components of a steel wire rope

Steel wire rope consists of three basic components:

1. Wires.
2. Strands, formed by wires, laid helically around a core.

3. Core, or centre, which can be of fibre (CF) or steel (CS).

These three elements when optimally combined, due to their complexity and variation, can produce composite cables possessing rather diverse purposes and characteristics [2].

2.3. Rope Properties

The important characteristics of wire rope relate to the number and size of the outer wires, and to a lesser extent, the inner wires. A small number of large outer wires results in better resistance to wear and corrosion. A large number of small wires results in better flexibility and resistance to fatigue.

1. The Wires
The wires are the basic units of construction of a steel cable, manufactured according to different specifications depending on the intended end use of the cable. The wire can be defined as the product of a uniform solid section along its entire length, obtained by cold deformation of the wire, and that can be rolled/twisted in turns, and arranged or not.

The deformation can be done by drawing (cold drawing without removal of material) of wire through a die, or by lowering the low-pressure passage between rollers connected or not, followed by the rolled product. The cross-section is usually circular, but can be oval, rectangular, square, hexagonal, octagonal or of convex shape.

2. The Cable Core
The main function of the cable core is to serve as a base, or central axis, where the strands will be rolled up to maintain the roundness of the cable and retain the strands so placed in their correct position during use. Furthermore, the core supports the pressure of the strands and maintains correct distance or space between them. The composition of the core will have an effect on cable performance and the use to which the cable is intended.

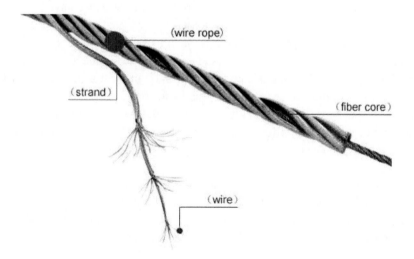

Figure 1. Basic components of a wire rope.

2.1. Types of Cores

There are two main types of cores:

- Fiber core (FC).

This centre is made of either natural fibres or polypropylene and similar other material, with greater elasticity than the independent wire rope core (IWRC). Polypropylene is standard, but either natural sisal (or hemp) fibre or some other man-made fibres are also available on special request.

- IWRC.

This is, in fact, an independent wire rope with strands and a core, and is called an IWRC. Most wire ropes made with steel core use an IWRC.

1. The Strand

The strand is one of the constituent elements of the cable, comprising a number of wires with a specified pattern according to their construction, which are helically wound around a central core in one or several layers. The inner layers assist in holding the outer wires, so that every set can slide freely, and especially when the cable has to bend. Each number and arrangement of the wires are called a construction, and are manufactured in one operation with all wires twisted in the same direction together in parallel operations. This eliminates the crossing and abrasion of the wires in the inner layers, which could weaken the cable and reduce its life, with a likelihood of failing without warning. The strands provide all the tensile strength in a fibre core wire, and over 90% of the resistance wire in a typical six-strand rope with a separate cable core. The strands are compacted during the action of twisting, thereby obtaining a larger metallic area and, therefore, greater resistance to breakage for the same nominal diameter, and a greater contact surface area with outer wires of pulleys and drums, thus giving a greater abrasion resistance, and therefore, less wear on the pulleys and drums. Cords offer higher crush strength and reduced internal vibrations as we have seen, as they can be of various types: steel, plastic or fibre. The number of strands, and the construction of the wire rope, determine the classification. Features, such as fatigue and abrasion resistance, are directly affected by the design of the laces. Features like fatigue strength and abrasion strength are directly affected by the design of the strands. In most of the strands with two or more layers of wire to hold the inner layers, this is so that the outer wires can freely self-adjust and slide, especially when the cable needs to bend. As a general rule, a cable having strands made with fewer wires will be more abrasion resistant and resistant to fatigue, much less than a cable of the same size and made of many more smaller wire strands.

3.1. The Wire Rope

This is the final product that is formed by several strands which are rolled helicoidally around a core. The wire rope is identified by the number of strands and by the number of wires in each strand, and the kind of core. The fundamental designation can be completed with details of:

1. Disposition for the wires.
2. Direction of wiring.
3. Disposition of strands.
4. Kind of wires.
5. Tension and internal balance.

3.2. Wire Rope Classifications

3.2.1. Based on the Type of Layer

- Single layer: The most common example of the single-layer construction is a seven-wire strand. It has a single-wire centre with six wires of equal diameter wound around it.

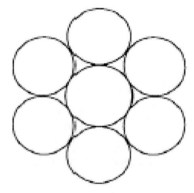

Figure 2. Single Layer wire rope.

- Seale: This construction has two layers of wires wound around a centre, with the same number of wires in each layer. All the wires in each layer are of equal diameter. The strand is so designed, that the larger-diameter outer wires rest in the valleys between the smaller-diameter inner wires. Example: 19 Seale (1-9-9) strand.

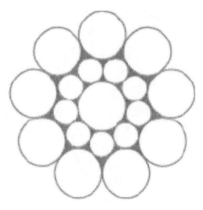

Figure 3. Seale (1-9-9).

- Filler wire: This construction has two layers of uniform-size wire wound around a centre, with the inner layer having half the number of wires as in the outer layer. Small filler wires, equal in number to the inner layer, are laid in the valleys of the inner layer. Example: 25 Filler Wire (1-6-6f-12) strand.

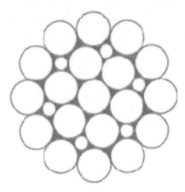

Figure 4. Filler (1-6-6f-12).

- Warrington: This construction has two layers of wires wound around a centre, with a uniform diameter of wires in the inner layer, and two different diameters of wire, alternating large and small, in the outer layer. The larger of the outer layer wires rest in the valleys, and the smaller ones rest on the crowns of the inner layer. Example: 19 Warrington (1-6-(6+6)), based on the nominal number of wires in each strand.

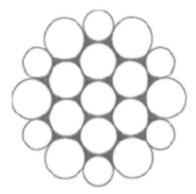

Figure 5. Warrington (1-6-(6+6)).

3.2.2. Another Definition of the Wire Rope
There is another definition of the wire rope. These wires can be classified according to the following criteria:

1. In terms of their construction process and heat treatments.
2. The sectional shape of the wires.
3. The type of steel.
4. The nature of the surface of the wire.
5. The position of the wires in the cable.
6. The function of the wires in the cable[4].

4. Definition and Classification of Wire Ropes Based on EN 12385-2: 2008 [5]

Explanation of symbols used

Figure 6. Example of wire rope section.

A: Rope, nominal diameter in mm
B: Rope construction

- The number of wires per strand may vary from 3 to 91, with the majority of wire ropes falling into the 7-wire, 19-wire, or 37-wire strand categories.

C: Construction and lay direction

- The Lay of a wire rope is simply a description of the way wires and strands are placed during construction. Right lay and left lay refers to the direction of the strands. Right lay means that the strands pass from left to right across the rope. Left lay means just the opposite: strands pass from right to left. Regular lay and lang lay describes the way wires are placed within each strand. Regular lay means that wires in the strands are laid opposite in direction to the lay of the strands. Lang lay means that wires are laid in the same direction as the lay of the strands. Most of the wire ropes used today are right lay and regular lay. This specification has the widest range of applications and meets the requirements of most equipment.

Table 1. Construction and lay direction of wire ropes

Lay	Definition	Characteristics
Regular Lay	Most common lay in which the wires wind in one direction and the strands in the opposite direction (right lay shown).	Less likely to kink and untwist; easier to handle; more crush resistant than lang lay.
Lang Lay	Wires in strand and strands of rope wind in the same direction (right lay shown).	Increased resistance to abrasion; greater flexibility and fatigue resistance than regular lay; will kink and untwist.
Right Lay	Strands wound to the right around the core (regular lay shown).	The most common construction.
Left Lay	Strands wound to the left around the core (regular lay shown).	Used in a few special situations, e.g., cable tool drilling line.
Alternate Lay	Alternate strands of right regular lay and right lang lay.	Combines the best features of regular and lang lay for boom hoists or winch lines.
Strands → Wires →		

- Single lay strand:

example for strand construction: 7 means (1-6)

- Seale (S) parallel lay:

example for strand construction: 19S means (1-9-9)

- Warrington (W) parallel lay:

example for strand construction: 19W means (1-6-6+6)

- Filler (F) parallel lay:

example for strand construction: 21F means (1-5-5F-10)

25F means (1-6-6F-12)

- Combined (Warrington Seale (WS)) parallel lay:

example for strand construction: 31WS means (1-6-6+6-12)

D: Construction of core

Single-layer rope with fibre core FC

- NFC: natural fibre core
- SFC: synthetic fiber core

Single-layer rope with steel core WC

- WSC: wire strand core
- IWRC: independent wire rope core

Rope with parallel lay

- PWRC: parallel wire rope centre

E: Nominal tensile grade of wires in N/mm^2

F: Surface finish of wires
- U: bright
- B: zinc coated (class B)

G: Type and direction of lay

- z: right lay (strand)
- s: left lay (strand)
- Z: right lay (rope)
- S: left lay (rope)
- sZ: regular lay, right-hand
- zS: regular lay, left-hand
- zZ: lang lay, right hand

3. THE MAIN PURPOSES OF INSPECTION OF THE CABLES

There are several purposes served in the inspection of the cables and devices, of which we highlight the following:

- Inspection shows the condition of the cable and indicates the need for replacement or not.
- The inspection may indicate whether the cable is being used correctly.
- The inspection may indicate if the cable type is appropriate.
- Inspection of cables allows to detect failures and indications/tell-tale marks on devices, that can result in premature wear of the cable [3].

The level and frequency of inspection depend on potential risks for the staff and the machinery.

4. KINDS OF INSPECTIONS

- Survey of supply conditions before entry into service
  - Before entry into service, and even before installation of the wire rope on the machine, in order to verify that these are correct and appropriate, it must be determined from the following factors and other characteristics of the cable.
  - Verify that suitable supply conditions have been met.
  - Cable diameter.
  - Mass per metre of cable.

- Quality coating (if the wires are galvanised).
- Tensile strength.
- Diameter of the wires.
- Survey of wire rope once commissioned into service
  - Once the wire ropes are in service, three types of inspections are set according to frequency: daily inspections; periodic examination; special examinations.

5. ITEMS THAT MUST BE INSPECTED

It is visually impossible to examine the cable along its entire length. Therefore, it is important to choose the areas that are most prone to wear and tear. For cables that are rolled on a drum, or pass over pulleys or rollers, it is recommended to examine the parts that fit in the grooves of the pulleys at the time of load application. These are parts that are subjected to shock at the stopping time, and areas particularly exposed to the weather. The fixing points at the ends of the wire rope. The wire rope areas which are subject to repetitive work [9].

6. CAUSES OF DAMAGE TO THE WIRE ROPES

Usually, the damage to wire ropes is caused by a combination of several factors which result in a so-called cumulative effect, whose final result is the discarding of the wire. All wire ropes in permanent service will eventually fail. Its causes for failure depend on the condition under which they operate. Rope life can be reduced due to a number of factors. The integrity of wire ropes deteriorates with time, and there are many other causes, like corrosion, fatigue and abrasion, which may finally break the ropes. As can be seen, there are many causes that lead to damage of wire ropes on board, these can be categorised into different types.

7. CRITERIA FOR REPLACEMENT BASED ON ISO 4309

There are a number of criteria for rejection, and they all agree on the factors described below. All these criteria have to be examined separately, but the combination of certain alterations in certain areas may result in a cumulative effect that must be taken into account by the person competent to decide on replacing or commissioning of the cable. The operational safety of the cables in service is based on the following criteria [6]:

7.1. Wire Protrusion

In wire protrusion, certain wires or groups of wires rise up, on the side of the rope opposite to the sheave groove, in the form of loops. A rope with wire protrusion should be immediately discarded.

7.2. Core Protrusion – Single-Layer Rope

This feature is a special type of basket or lantern deformation in which the rope imbalance is indicated by protrusion of the core (or centre of the rope, in the case of a rotation-resistant rope) between the outer strands, or protrusion of an outer strand of the rope or strand from the core. A rope with core or strand protrusion/distortion should be immediately discarded.

Figure 7. Wire protrusion.

Figure 8. Core protrusion single-layer rope.

7.3. Local Reduction in Rope Diameter (Sunken Strand)

Reduction of rope diameter resulting from deterioration of the core can be caused by:

1. Internal wear and wire indentation.
2. Internal wear caused by friction between individual strands and wires in the rope, particularly when it is subjected to bending.
3. Deterioration of a fibre core.
4. Fracture of a steel core.
5. Fracture of internal layers in a rotation-resistant rope.

If these factors cause the actual rope diameter to decrease by 3% of the nominal rope diameter for rotation-resistant ropes, or by 10% for other ropes, the rope should be discarded even if no broken wires are visible.

7.4. Strand Protrusion/Distortion

This feature is a special type of basket or lantern deformation in which the rope imbalance is indicated by protrusion of the core (or centre of the rope, in the case of a rotation-resistant rope) between the outer strands, or protrusion of an outer strand of the rope or strand from the core. A rope with core or strand protrusion/distortion should be immediately discarded.

Figure 9. Local Reduction in rope diameter.

Figure 10. Strand protrusion.

7.5. Flattened Portion

Flattened portions of a rope which pass through a sheave will quickly deteriorate, exhibiting broken wires, and may damage the sheave. In such cases, the rope should be discarded immediately. Flattened portions of rope in standing rigging can be exposed to accelerated corrosion, and must be subjected to inspection at a prescribed shortened frequency if retained in service.

7.6. Kink-Positive or Negative

A kink or tightened loop is a deformation created by a loop in the rope which has been tightened without allowing for rotation about its axis. Imbalance of lay length occurs, which will cause excessive wear, and in severe cases, the rope will be so distorted that it will have only a small proportion of its strength remaining. Rope with a kink or tightened loop should be immediately discarded.

Figure 11. Flattened portion.

Figure 12. Kink-positive.

Figure 13. Kink-negative.

7.7. Deformation

Visible distortion of the rope from its normal shape is termed deformation, and can create a change at the position of deformation which results in uneven stress distribution in the rope.

7.8. Waviness

Waviness is a deformation in which the longitudinal axis of the wire rope takes the shape of a helix under either a loaded or unloaded condition. While not necessarily resulting in any loss of strength, such a deformation, if severe, can transmit a pulsation resulting in irregular rope drive. After prolonged working, this will give rise to wear and wire breaks. In the case of waviness, the wire rope should be discarded if, under any load condition on a straight portion of rope that does not bend around a sheave or drum, the following condition is met:

d_1 4 $d/3$

or, on a portion of rope that bends around a sheave or drum, the following condition is met:

d_1 1.1d

where;

d is the nominal diameter of the rope;
d_1 is the diameter corresponding to the envelope of the deformed rope.

Figure 14. Waviness.

Figure 15. Basket deformation.

7.9. Basket Deformation

Basket or lantern deformation, also called birdcage, is a result of a difference in length between the rope core and the outer layer of strands. Different mechanisms can produce this deformation.

If, for example, a rope is running over a sheave or onto the drum under a great fleet angle, it will touch the flange of the sheave or the drum groove first and then roll down into the bottom of the groove. This characteristic will unlay the outer layer of strands to a greater extent than the rope core, producing a difference in length between these rope elements.

When running over a "tight sheave", that is, a sheave with a groove radius that is too small, the wire rope will be compressed. This reduction in diameter will, at the same time, result in an increase in rope length. As the outer layer of strands will be compressed and lengthened to a greater extent than the rope core, this mechanism again will produce a difference in length between these rope elements.

7.10. External wear

Abrasion of the crown wires of outer strands in the rope results from rubbing contact under pressure, with the grooves in the sheaves and drums. The condition is particularly evident on moving ropes at points of sheave contact when the load is being accelerated or decelerated, and is revealed by flat surfaces on the outer wires.

Wear is promoted by lack of lubrication, or incorrect lubrication, and also by the presence of dust and grit. Wear reduces the strength of ropes by reducing the cross-sectional area of the steel strands. If, due to external wear, the actual rope diameter has decreased by 7% or more of the nominal rope diameter, the rope should be discarded even if no wire breaks are visible.

7.11. External Corrosion

Corrosion occurs particularly in marine and polluted industrial atmospheres. It will diminish the breaking strength of the rope by reducing the metallic cross-sectional area, and will accelerate fatigue by causing surface irregularities which lead to stress cracking. Severe corrosion can cause decreased elasticity of the rope. Corrosion of the outer wires can often be detected visually. Wire slackness due to corrosion attack/steel loss is good justification for immediately discarding the rope.

Figure 16. External wear.

Figure 17. External corrosion.

7.12. Internal Corrosion

This condition is more difficult to detect than external corrosion which frequently accompanies it, but the following indications can be recognised:

a) variation in rope diameter.

In locations where the rope bends around sheaves, a reduction in diameter usually occurs. However, in stationary ropes, it is not uncommon for an increase in diameter to occur due to the build-up of rust under the outer layer of strands.

b) loss of clearance between the strands in the outer layer of the rope, frequently combined with wire breaks between or within the strands.

If there is any indication of internal corrosion, the rope should be subjected to internal examination. Confirmation of severe internal corrosion is good justification for immediately discarding the rope.

7.13. Wire Breaks in Crown or Valley

Broken wires usually occur at the external surface. In the case of rotation-resistant ropes, there is the probability that the majority of broken wires will be found internally and are, thus, "non-visible" fractures. One valley break may indicate internal rope deterioration, requiring closer inspection of this section of rope. When two or more valley breaks are found in one lay length, the rope should be considered for discard. When establishing rejection criteria for rotation-resistant ropes, consideration shall be given to the rope construction, length of service and the manner in which the rope is being used. Particular attention shall be paid to any localised area which exhibits a dryness or denaturing of the lubrication.

Figure 18. Wire break at the crown of the strand.

Figure 19. Valley wire breaks.

7.14. Protrusion of Inner Rope of Rotation-Resistant Rope

This feature is a special type of basket or lantern deformation in which the rope imbalance is indicated by protrusion of the core (or centre of the rope, in the case of a rotation-resistant rope) between the outer strands, or protrusion of an outer strand of the rope or strand from the core. A rope with core or strand protrusion/distortion should be immediately discarded.

7.15. Local Increase in Rope Diameter due to Core Distortion

A local increase in rope diameter can occur, and might affect a relatively extensive length of the rope. This condition usually relates to a deformation of the core (in particular environments, a fibre core can swell up owing to the effect of moisture) and, consequently, creates imbalance in the outer strands, which become incorrectly oriented.

If this condition causes the actual rope diameter to increase by 5% or more, the rope should be immediately discarded.

7.16. Kink

A kink or tightened loop is a deformation created by a loop in the rope which has been tightened without allowing for rotation about its axis. Imbalance of lay length occurs, which will cause excessive wear, and in severe cases, the rope will be so distorted that it will have only a small proportion of its strength remaining. A rope with a kink or tightened loop should be immediately discarded.

Figure 20. Protrusion inner rope of rotation-resistant rope.

Figure 21. Local increase in rope diameter due to core distortion [8].

7.17. Flattened Portion

Flattened portions of rope which pass through a sheave will quickly deteriorate, exhibiting broken wires, and may damage the sheave. In such cases, the rope should be discarded immediately. Flattened portions of rope in standing rigging are liable to being exposed to accelerated corrosion, and must be subjected to inspection at a prescribed shortened frequency if retained in service.

Figure 22. Kink.

Figure 23. Flattened portion.

Figure 24. Broken wires at termination.

7.18. Broken Wires at Termination

Broken wires at, or adjacent to, the termination, even if few in number, are indicative of high stresses at this location, and can be caused by incorrect fitting of the termination. The cause of this deterioration must be investigated and, where possible, the termination shall be remade, shortening the rope, if sufficient length remains, for further use, failing which the rope should be immediately discarded.

7.19. Localised Grouping of Broken Wires

Where broken wires are very close together, constituting a localised grouping of such breaks, the rope should be discarded. If the grouping of such breaks occurs in a length less than 6d, or is concentrated in any one strand, it may be necessary to discard the rope.

Fracture of Strands
If a complete strand fracture occurs, the rope should be immediately discarded.

Decreased Elasticity
Under certain circumstances usually associated with the working environment, a rope can sustain a substantial decrease in elasticity and is, thus, unsafe for further use. Decreased elasticity is difficult to detect. If the examiner has any doubt, advice shall be obtained from a specialist in wire ropes. However, it is usually associated with the following:

a) reduction in rope diameter
b) elongation of the rope lay length
c) lack of clearance between individual wires and between strands, caused by the compression of the component parts against each other
d) appearance of fine brown powder between or within the strands
e) increased stiffness.

While no wire breaks may be visible, the wire rope will be noticeably stiffer to handle, and will certainly have a reduction in diameter greater than that related purely to wearing of individual wires. This condition can lead to abrupt failure under dynamic loading, and is sufficient justification for immediate discard.

Bends
Bends are angular deformations of the rope caused by external influence. A rope with a severe bend will suffer similar to flattened portions of rope.

Tensile Breaks
Tensile breaks show typical necked-down ends, one side of the break being coned, and the other cupped.

Wire Rope Damage on Pulley and Drums

The pulleys, or drums, can cause damage to the wire rope for several reasons, among which we can highlight the following:

7.20. Abrasion

The main cause of damage to the wire rope pulleys and drums is abrasion, therefore, it is vital that all the components are in proper working condition with the appropriate wire rope diameter. A pulley, or drum, can cause serious damage to a new wire rope, resulting in the premature replacement of the wire rope.

7.21. Fatigue Breaks

The fatigue life of a rope is determined by the combination of loads and the diameters of any sheaves and drums. Fatigue breaks may be accelerated by abrasion or peening, and are normally produced under conditions such as: continuous or reversed bending; vibration, particularly where wires and strands become locked up through lack of lubrication; corrosion; compression in badly worn sheave grooves or over the equaliser sheaves. If the conditions are linked to corrosion, failure of wires by fatigue is accelerated.

Figure 25. Abrasion on a drum.

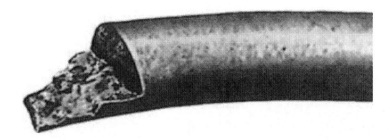

Figure 26. Fatigue breaks [9].

8. TECHNIQUES FOR INSPECTION OF WIRE ROPES

8.1. Visual Inspection

The most conventional inspection methods for wire ropes are the visual inspections, which are a simple yet useful method, in which the experts observe the surface and assess the rope condition empirically, for detecting a wide variety of external rope deteriorations. Very often, the rope is covered with grease, thus, many external and internal defects elude such an examination.

Another visual inspection tool is measurement of the rope diameter. Rope diameter measurements compare the original diameter with the current reading under like circumstances. A change in rope diameter indicates external and/or internal rope damage. Inevitably, many types of damage do not result in a change of rope diameter.

During a visual inspection for wire ropes, every single wire of the rope cannot be inspected, so a great percentage of the metallic wire area cannot be visually inspected at all, thus, evaluation of the inner failure, such as corrosion, cannot be carried out.

The only wires that can be visually inspected are the outer wires of the rope. These represent about 40% of the metallic cross-sectional area, but even these wires disappear inside the rope for about half their lengths, leaving only about 20% of the rope's cross-section accessible for a visual inspection [10].

With visual inspection, only a small percentage of the steel wire rope cross-section can be inspected; visual rope inspections must, therefore, be carried out with great care.

The actual result of a visual rope inspection is 20% for the visual inspection, and 80% by way of estimation.

Therefore, visual inspections have a limited value as the sole means of wire rope inspection. However, visual inspections are simple and do not require special instrumentation. When combined with the knowledge of an experienced rope examiner, visual inspection can provide a valuable supplementary tool for evaluating many forms of rope degradation.

Electromagnetic inspections

In order to also gain information about the remaining 80% of the steel wire rope cross-section, other techniques to detect failures in wire ropes have been developed. These are the electromagnetic method and other NTD non-destructive evaluations.

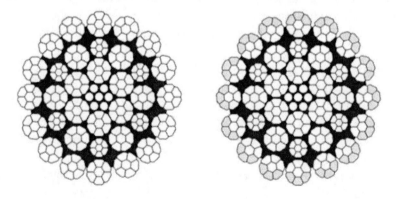

Figure 27. Representation of the visible area of a wire rope.

Because many ropes deteriorate internally without externally visible signs, their inspection solely by visual methods is unsafe. In other cases, visual inspection is possible but not practical. For example, mooring ropes are often covered with mud and marine growth. This makes their visual inspection difficult, if not impossible [11].

When visual inspections are impossible, rope operators usually resort to a statutory life policy as an alternative. The statutory life method requires rope retirement at prescribed fixed intervals, but this approach must be used extremely conservatively and, therefore, is not profitable. Yet, in spite of this very uneconomical practice, occasional rope failures still occur.

More dependable inspection methods, combined with a better understanding of degradation mechanisms and discard criteria, can notably increase wire rope life and human safety.

For example, the electromagnetic method used at regular intervals, can significantly increase the safety of installations using wire ropes.

The electromagnetic method is particularly effective when it is combined with visual examination and thorough understanding of rope deterioration modes.

8.2. Magnetic Inspection Operating Principle

Wire ropes have magnetic properties because they are made with drawn steel wires. When a wire rope is longitudinally saturated by a strong magnetic field, the magnetic flux in the rope is well proportional to the rope's metallic cross-sectional area.

Non-destructive testing of wire ropes can be provided with the wire rope tester. The wire rope test instruments are of the so-called dual-function localised flaws/loss of metallic cross-sectional area (LF/LMA) type, which means that they can reliably detect LFs, typically caused by external and internal broken wires and corrosion pitting, and accurately measure LMA due to external and internal corrosion and wear.

The wire rope test instrument operates on the principle of magnetic flux leakage (MFL). The magnetic head can be of different types, and its function is to saturate the rope. The sensors are the measuring system. Grease, paint, dirt or plastic sealing material, inspection speed does not affect the results of the inspection.

Any anomaly of the rope under test causes a change of the magnetic flux, such as change of the leakage flux around the rope.

The changes are measured by magneto-sensitive sensors (Hall sensors, most often), or by inductive coils.

There are a number of magnetic flaw detectors manufactured by various companies and offered on the world market.

These modern instruments demonstrate testing results close to each other. Nevertheless, they differ by specifications which are important for users, for instance, by mass and dimensions; by method of testing data storing, processing and presenting; by types and range of ropes under test dimensions.

The basic components of the standard equipment used in the electromagnetic method is indicated in the figure:

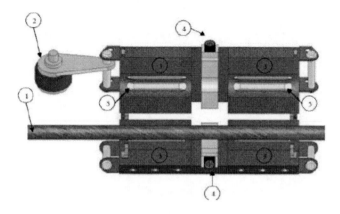

Figure 28. Basic elements of wire rope magnetic testing equipment.

1. Wire rope to be inspected.
2. Speed sensor, which measures the relative velocity between the cable and equipment. This sensor determines the speed at which the assay is performed, and the distance from the initial reference point on the cable or installation from which the defects can be located.
3. Permanent magnets.
4. Sensors LF/LMA.

The LMA signal gives a quantitative measure of the loss of metallic cross-sectional area of the wire rope caused by corrosion pitting, broken wires and other anomalies.

The LF signal can pinpoint the location of a wide variety of flaws, such as broken wires and corrosion pitting. After flaws have been located by using the LF signal, their actual nature can be ascertained by analysing the corresponding section of the LMA signal and/or by performing a visual wire rope inspection.

1. Skids or slides. These are intended to limit the relative movement of wire sideways, and in order to stabilise it to prevent false signals in the diagram, or interference caused by the undesired movement of the wire rope, commonly called noise.

Figure 29. Measurement head.

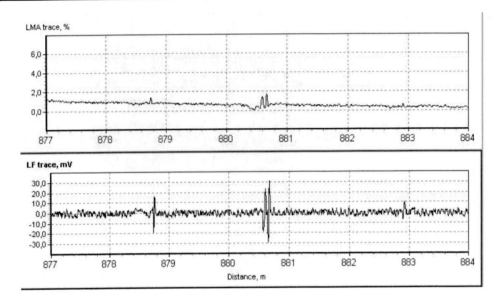

Figure 30. Data chart of a rope.

Image of an actual head through which is passed the steel cable that, once it is in operation for the inspection process, the permanent magnets induce an electromagnetic field in the cable. The sensors are placed on the head, and LF/LMA are responsible for detecting changes in the magnetic field. These variations in the magnetic field are recorded, analysed and evaluated in order to identify possible defects in the power cable and, thus, determine when the cable should be taken out of use. This type of non-destructive testing is to pass the cable through the inspection team or head, which generates, through appropriate sensors, an electromagnetic field through the wire. Changes in the magnetic field are recorded using a recorder, to finally be interpreted, and to pinpoint where to find one or more broken wires inside the cable, and to evaluate other possible failures, such as abrasion and internal corrosion. The signals indications are represented graphically as shown in the figures.

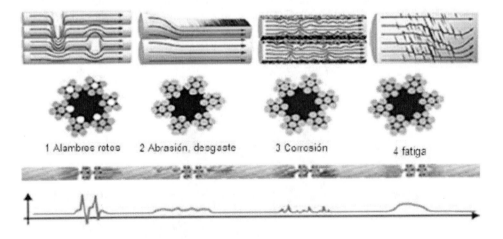

Figure 31. Indications represented graphically.

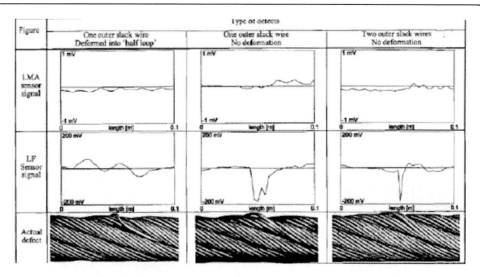

Figure 32. Indications represented graphically.

These rope testers can detect rope deterioration at the earliest stages. Therefore, wire rope users can employ them as an effective preventative maintenance tool.

Using the magnetic rope tester, the early detection of corrosion allows immediate corrective action through improved lubrication. Accelerating wear and interstrand nicking can indicate a need to reline sheaves to stop further degradation.

Careful inspections can monitor the development of local damage at the crossover points of the rope on a winch drum. In this manner, the operator can determine the optimum time for repositioning the rope on the drum.

CONCLUSION

- To prevent a disaster, we must inspect wire ropes periodically by using a combination of different methods, like visual inspection, NDT inspection and magnetic inspection, to evaluate the degree of deterioration.
- The criteria for replacement of wire ropes must allow a safety use of wire rope until operation is removed.
- The wire rope inspection is a check on the condition of the wire rope, to identify a malfunction or deterioration that could affect its utility, therefore, it is absolutely necessary to perform a deeper analysis. Hence, visual inspection is insufficient
- Current developments allow the electromagnetic method to be used quickly and more reliably than visual inspection.
- The NDT by magnetic method inspection of wire ropes is commonly used in other industries throughout the world, but it is not ordinarily used by ships and offshore operators.
- Rope operators usually resort to a statutory life policy as an alternative. The statutory life method requires rope retirement at prescribed fixed intervals, but this approach must be used extremely conservatively and, therefore, is not profitable. Yet, in spite of this very uneconomical practice, occasional rope failures still occur, which can be

avoided using more dependable inspection methods, such as the electromagnetic method combined with visual inspection and a better understanding of degradation mechanisms and discard criteria, which can notably increase wire rope life and human safety.

REFERENCES

[1] US Army Corps of Engineers. (2004). *Wire rope selection criteria for gate operating devices*. 1110-2-32.

[2] *Wire Rope User's Handbook*. UNION. www.wirecoworldgroup.com

[3] UNE-EN 13852-1:2005. (2005). Cranes - Offshore cranes - Part 1: General purpose offshore cranes.

[4] Casar. *Wire Rope Inspection and Examination*. www.casar.de (Accessed December 2012).

[5] UNE – EN 12385-2:2004. (2004). *Steel wire ropes - Safety* - Part 2: Definitions, designation and classification.

[6] ISO 4309: 2010. (2010). *Wire ropes - Care and maintenance, inspection and discard*.

[7] ASTM E570-09 *Standard Practice for Flux Leakage Examination of Ferromagnetic Steel Tubular Products*.

[8] Secul, P., Aguirre, F., Véliz, G. (2007). *END de Cables de Acero: Ensayos Magnetográficos*. IV Conferencia Panamericana de END Buenos Aires.

[9] Weischedel, H. R. Crane Wire Rope Damage and Nondestructive Inspection Methods. NDT Technologies, Inc. http://www.ndttech.com /Papers/FluxLeakageWireRopes.pdf. (Accessed December 2012).

[10] *Non destructive Inspection of Steel Wire Ropes*. INTRON. www.intron.ru. (Accessed December 2012).

[11] Sukhorukov, V. V., Kotelnikov, V. S., Zhukov, V.G. and Khudoshin, A. A. (2003).Importance of rope NDT for safe lifting of loading cranes. ODN 0750 Intron Plus Ltd. OIPEEC Technical Meeting Lenzburg.

SECTION 2: SHIPS AND ENVIRONMENTAL IMPACT

SUBSECTION 3: THE INTERNATIONAL MARITIME ORGANIZATION (IMO)

In: Ships and Shipbuilding
Editor: José A. Orosa

ISBN: 978-1-62618-787-0
© 2013 Nova Science Publishers, Inc.

Chapter 7

RESEARCH ABOUT THE NEW IMO CONVENTION

Rebeca Bouzón, Ángel M. Costa, A. De Miguel Catoira, J. Romero Gómez and M. Romero Gómez*
Department of Energy and Marine Propulsion
University of a Coruña, Spain

ABSTRACT

The dismantling and/or scrapping of sea-going vessels that have outlived their operational life is one of the most intricate and problematic environmental challenges confronting the maritime sector today. Numerous publications that are available, generally relate to maintenance and design technologies and the construction of ships, but in this chapter, we will discuss at length, the dismantling of ships. The International Maritime Organization (IMO) in 1958, had already produced several important international conventions, including the International Convention for the Safety of Life at Sea in 1948, and the International Convention for the Prevention of Pollution of the Sea by Oil in 1954, and several treaties relating to load lines and prevention of collisions between ships at sea. The IMO is mandated with ensuring the upkeep of most of these conventions. It is also entrusted with the task of developing new agreements as and when the circumstances so warrant. The creation of the IMO coincided with a period of profound change in the world of shipping, so that, from the beginning.

The IMO had to take great efforts to develop new agreements, and ensure that existing instruments should maintain the same pace with day-to-day technologies evolving in the shipping industry. Currently, the IMO is charged with 35 international conventions and agreements, and it has brought out numerous protocols and amendments. In this paper, we will talk about the new convention of the IMO.

INTRODUCTION

In Hong Kong, in May 2009, The International Maritime Organization (IMO) in an international convention for the safe and environmentally sound recycling of ships, after

---

* Email: jaorosa@udc.es.

several years of work which involved the IMO and International Labour Organization (ILO), adopted the unified part of the Basel Convention, and various other non-governmental organisations [1].

In 2007, the Commission of the European Community stated that the dismantling of ships was sustainable from their point of view, but was very costly in terms of both human health and the environment. This problem had long worried the various organisations, with increasing pressures calling for recycling to be made safer and environmentally friendly as well [2]. This paper will also address the issue of how previous practice of scrapping vessels and several accidents occurred simultaneously.

On the other hand, we will also discuss the first attempt to address this problem through the Basel Convention, and how recycling will eventually improve the development of the International Convention for the Safe and Environmentally Sound Recycling of Ships, which will hereafter be called the "Hong Kong Convention".

The entry into force of this Agreement will occur twenty-four months after it has been approved by at least 15 countries representing at least 40% of the gross tonnage of the world merchant fleet, and that the total annual volume of these vessels scrapped by a country during the previous 10 years represents at least 3% of the gross tonnage of its merchant fleet [1].

In this chapter, we discuss a diplomatic conference convened by the IMO which was developed in Hong Kong between 11 and 15 May 2009, which put the finishing touches to, and formally adopted the so-called Hong Kong Convention.

The meeting was attended by delegates and observers from 65 countries: the UN agency for environment, the ILO, the European Commission and eight NGOs. This new Convention addressed all the problems concerning the dismantling and recycling of ships, and especially to two aspects:

- The treatment of the potentially hazardous substances or contaminants that may contain hazardous material, such as asbestos, mercury and other heavy metals, hydrocarbons, substances damaging to the ozone layer, and so on.
- The conditions of human employment and the environment in which they engage in the activities of dismantling and recycling.

Their deliberations have taken three years of hard work; they have participated in the IMO secretariat, its member states and several non-governmental organisations concerned, in collaboration with the ILO and the states which are parties to the Basel Convention on the control of transboundary movements of hazardous waste.

The IMO had begun its work on recycling much before this time, and as early as 2003 released recommendatory guidelines on this matter. Ship dismantling is the process of disarming the obsolete structure of a ship. This operation is carried out in a shipyard, or boatyard, with the vessel disarmed, and includes a variety of activities, from removing all gear and equipment to cutting down and recycling the ship's infrastructure. Ship-breaking is a difficult process due to the structural complexity of the ships and the many issues that address environment, safety and health.

1. Risks of Dismantling

Ship dismantling exposes workers to a wide range of hazards. These hazards are as follows:

- Asbestos-coated coverings, mastics applied under the insulation, fabric above the insulation, cables, brake linings, pipe insulation, hulls, adhesives, gaskets and pipe connections of valves and containers.
- Polyclorobiphenyls (polychlorinated biphenyls – PCBs) in products, such as rubber hoses, plastic foam insulation, cables, silver paint, painting in habitable zones, felt under soundproofing plates, the plates above the hull bottom and hull painting of steel primaries.
- Lead from lead paint or chromate of lead ballast, batteries, generators and engine components.
- Hazardous materials and chemicals, including heavy metals, ships' transducers, ballast and paint coatings, fluorescent light tubes with mercury, thermometers, electrical switches, light fixtures, smoke detectors and indicators of tank levels, and chlorofluorocarbons (CFCs) in autonomous systems, such as cooling water sources and small refrigerated freezer units.
- Excessive noise associated with grinding, hammering, metal cutting and other activities.
- Fire isolates from flames, mates, linings and residual fuel and lubricants and other flammable liquids.

1.1. Work Activities at Risk

- Access to areas in confined spaces and atmosphere, closed and otherwise dangerous.
- Removal of paint. Cut and scrap metal.
- Activities with forklifts.
- Work carried out on elevated surfaces, particularly near edges and covered openings.
- Withdrawal of ballast and bilge water.
- Removal of oil and fuel, and tank cleaning.
- Removal and disposal of machinery of ships.
- Operations that require the use of cranes, gears and material-handling equipment.
- Welding and cutting operations, and use of compressed gas.
- Activities that require the use of scaffolding, ladders and operating services, and other working conditions involving risk.
- Inadequate training of workers.
- Personal protective equipment, absent or inappropriate, as show in Figure 1.
- Inadequate fire protection measures.
- Lack of emergency response, rescue services, and personnel and relief.

Figure 1. Workers cutting plates without adequate protective gear.

2. New Convention on Ship Recycling

The dismantling of ships is an activity that is carried out by the method of "beaching" (Figure 2). This method involves beaching the ship on to the beach where it will be scrapped by personnel who have neither the knowledge nor adequate preparation for this type of activity, in addition to not having the necessary protection against hazardous substances.

Figure 2. Ships for dismantling beached on a beach in Bangladesh.

The beaches where this activity takes place today, can be seen in countries like India, Bangladesh and Pakistan, where the largest number of recycling yards are in operation [3].

However, this has not always been the case, as initially, scrapping or dismantling of ships was an event that went parallel with ship construction, the most important being in Japan, Spain, Italy and South Korea [4].

So what happens in the dismantling of ships that have travelled to these countries? The answer lies in the price you can pay for a vessel to be scrapped. This price is linked to supply and demand, that is, the number of ships that need scrap steel and what is available to the industry. This price is calculated per ton of ship displacement, TDR [5] (ship displacement is the value expressed in metric tons representing the weight of a ship without cargo, fuel, lubricants, ballast, fresh water, provisions or crew) [6].

It is these countries that can offer a higher purchase price for the vessels to be recycled, because they have cheap labour, and lack of safety standards is not an issue: occupational safety, environmental degradation, and so on [5], which leads to the assumed costs being very low.

In response to these deficiencies, the risk of accidents in shipyards is high because, as mentioned above, there is a lack of appropriate equipment and training necessary to perform this work. Moreover, workers in the shipyards where ships are dismantled are also exposed to chronic diseases when handling and inhaling toxic substances [3].

According to the controlling organisation in Chittagong (Bangladesh), in the last decade, of the hundreds of people working in the yard, 70 have died, have been injured or have been poisoned, with an average of one worker dead and one injured every week day [7]. To mention some of the most recent accidents:

- In August 2009, six workers died at Alang (India) due to burns sustained during work performed in the engine room of the ships that were scrapped, as a result of not having taken the necessary precautions to ensure a gas-free area.
- In October 2010, also in Alang, an employee died due to waste material falling into the open part of the tanker vessel.
- In November 2011, in Mumbai (India), three workers suffered serious burns caused by a fire when cutting into the engine room of the ship being dismantled [8].
- In October 2012, in Alang, six workers were killed because of a fire that broke out while working in the pump room of a tanker [9].

3. HISTORY

During the forty-second session of the Committee for the Protection of the Marine Environment (MEPC 42) in November 1998, the issue of ship recycling was addressed first. Norway was urged upon, that environmental problems resulting from the dismantling of ships should be dealt with in the following sessions of the MEPC.

In the forty-third session (MEPC 43) of the Committee in July 1999, Norway introduced a proposal to add the issue of ship recycling to subsequent sessions. This proposal was accepted, and also that the IMO should play an important role in working with other parties, such as the Basel Convention, which was prepared to start treating the same subject in the fifth Conference of the Parties (COP 5) to be held in December 1999 [4].

Member states of the European Union (EU) are the owners of most of the world fleets, and are, therefore, to a large extent responsible for the way in which they carry out the dismantling of ships, as first-control means are sought in applying the Basel Convention, however, there has been a continuous violation of the rules of this Convention [3].

The Basel Convention on the Control of Transboundary Movements of Hazardous Wastes and their Disposal was adopted on March 22, 1989 in Basel (Switzerland). The Convention entered into force on May 5, 1992. Today, the membership consists of the same 178 countries.

This Convention aims to protect the health of the workers and the environment from the harmful effects of hazardous wastes, the Convention's objectives are:

- First, reduce the waste generation and its disposal, and ensure that these are conducted in an environmentally safe manner.
- Avoid or reduce transboundary movement of hazardous wastes, except in cases in which the safety of the environment is ensured.
- Regular cases where this movement is permitted [10].

At the Third Conference of the parties (the supreme decision-making organ of the Basel Convention) in 1995 (COP3), approved the Ban Amendment to the effect that, as of 1998, prohibiting the transfer of hazardous waste from OECD countries to non-OECD (Organization for Economic Cooperation and Development) countries, this prohibition was adopted as protection to underdeveloped countries that lack the means to prevent the illegal entry of toxic waste into its territories, or means to process them safely.

This amendment has not entered into force due to lack of ratifying the minimum number of requirements. Spain and other countries of the EU actually applied the export ban as far back as 1998. The countries that opposed the ban argued that there were countries outside the OECD where hazardous wastes are a source of raw materials, and they can manage them appropriately. They also pointed out that this amendment does not solve the problem of the transfer of waste between countries outside the OECD, or receiving internally processed hazardous substances produced in these countries. This delay has been due to the interpretation of Article 17 of the Basel Convention, which stipulates the minimum number of parties that have to approve an amendment for it to enter into force. Such ratification required three-fourths of the states parties, but there could be differences between the calculation performed and the number of countries that are party to the Convention in 1995, and the current number of countries [11]. In the 10th Conference of the Parties (COP10), they reached a consensus to launch the amendment. It clarifies the interpretation of Article 17 of the Basel Convention, and dictates that the Ban Amendment is to enter into force when ratified by 17 other parties, (71 parties have already ratified it) [12]. According to Article 2 of the Basel Convention, the term, wastes, are substances or objects which are disposed of, intended for process, or are required to be in accordance with national legislation. "Transboundary movement" means any movement of hazardous wastes or other wastes from an area under the national jurisdiction of one state to an area under the national jurisdiction of another state. The movement has to affect at least two states. And, by "sound management of hazardous wastes or other wastes" means taking all practicable steps to ensure that hazardous wastes and other wastes are managed in a manner that protects the environment and human health against the harmful effects that may result from handling such wastes [10]. Because almost all ships have dangerous substances, such as oils, hydrocarbons, heavy metals and so on when they are to be scrapped, these can be considered as hazardous waste. According to this, the vessels belonging to an EU state should be scrapped in OECD countries, hence, the Article in the Basel Convention. However, as stated above, vessels break the law continually, as in practice it is very difficult to know when an owner is getting rid of a vessel, as usually, these ships sail under their own power up to the point of breaking, and thus, cannot be treated as waste, and this is, therefore, very challenging to implement once the vessel leaves EU waters. That is why the parties to the Basel Convention requested the IMO to establish mandatory requirements for ship recycling. In 2006, the IMO introduced a draft convention which was adopted in 2009 as the Hong Kong Convention for the Safe and Environmentally Sound Recycling of Ships [3].

4. DEVELOPMENT OF THE HONG KONG CONVENTION

The new Convention is structured as follows:

- 21 articles, setting out the main legal mechanisms.
- 25 regulations, which contain technical requirements divided into four chapters:

  1. General (Regulations 1–3)
  2. Requirements for ships (Regulations 4–14)
  3. Requirements for ship recycling facilities (Regulations 15–23)
  4. Notification (Regulations 24–25)

- 7 appendices, lists of hazardous materials, forms for certificates, and so on.

Under Article 3, the Hong Kong Convention shall apply to:

- Vessels flying the flag of a state party.
- Recycling facilities operating under the jurisdiction of a state party.

This Convention shall not apply to:

- Warships, non-commercial ships operated by a state party and ships of less than 500 GT, provided that the parties ensure the adoption of appropriate measures for ships to act consciously with the Convention [13].

Article 4 of this Convention states that each state party shall enforce controls relating to:

- Vessels possessing its flag or operating under its authority.
- Recycling facilities within its jurisdiction.

Both must meet the requirements of the Convention [1]. In summary, the rules of the new Convention on the one hand refer to the ship, its construction, and preparation for safe recycling and dismantling facilities to ensure safe and sustainable operations [14]. All countries that have ratified this Convention shall ensure that hazardous materials listed in Appendix 1 of the same Article, are not used in their yards or on their ships. All vessels shall be fitted with Inventory of Hazardous Materials (IHM) with the amount and type of hazardous material they have. They must also undergo a renewal to verify that the IHM continues to adhere to the Hong Kong Convention, issuing a certificate called International Certificate on Inventory of Hazardous Materials (ICIHM), with a maximum validity of five years. These vessels can only be recycled at a facility of the member states. A ship that is prepared for recycling must meet the following requirements:

- Selecting an approved recycling facility in a member state which is capable of dealing with the types and quantities of hazardous materials on board.
- Providing a copy of the IHM and ICIHM and other important information relating to the facility, and to develop a recycling plan.
- Shall notify the management of the flag state that intends to recycle the ship.
- The plant will provide the Ship Recycling Plan upon completion of the job, and issue the international Ready for Recycling Certificate.

In addition, the facilities must also meet certain requirements, such as:

- Preparation of Ship Recycling Facility Plan to be covered by worker safety and training, protection of human health and the environment, and emergency preparedness. Furthermore, it should only accept ships complying with the Hong Kong Convention [15].

After completing the design as per the Hong Kong Convention, member states of the Basel Convention discussed the similarities between the two Conventions on the level of control and enforcement, but failed to reach an arrangement, as for IMO and the EU, the assessment was good [3]. However, other organisations believe that the Convention is a setback, according to the same, since the Hong Kong Convention does not prevent the transboundary movement of hazardous substances. The Basel Convention covers both recycling and final disposal, ensures that hazardous materials are properly treated, and has set out guidelines to be followed after scrapping; however, the Hong Kong Convention only applies to the yard, but this can cause problems after removal of the ship's hazardous materials, as they are no longer covered by that Agreement, and may be used in a reckless manner, or be emptied in the same country that they enter [16]. The entry into force of this Convention will occur twenty-four months after it has been approved by at least 15 countries representing at least 40 % of the gross tonnage of the world's merchant fleet, and that the total annual volume of scrapped vessels of these countries during the previous 10 years represents at least 3 % of the gross tonnage of its merchant fleet. In regard to the maritime sector, the most representative international associations are:

BIMCO: Baltic and International Maritime Council.
INTERCARGO: International Association of Dry Cargo Ship owners.
INTERTANKO: International Association of Independent Tanker Owners.
ICS: International Chamber of Shipping.
ITOPF: International Tanker Owners Pollution Federation.
ITF: International Transport Workers Federation.
OCIMF: Oil Companies International Marine Forum.

They decided to develop a Code of "Good Practices", based on ship dismantling, until the entry into force of the IMO Convention. Some of these include:

- The recommendation of a recycling contract so that the parties will undertake recycling taking into account the environment, safety and health of the people involved.
- Promote countries with dismantling facilities to have a "Certificate of Approval" for their facilities regarding safety measures and environmental protection.
- Incentivise facilities to make public the mechanism used to treat hazardous wastes, such as asbestos, halons, oils, and so on [4].

CONCLUSION

The dismantling of ships is an important and necessary task, and if it is not carried out, could endanger the safety of maritime traffic as a result of continuing to sail old ships that are no longer safe and seaworthy.

However, this activity has become a dangerous process after moving to Asian countries like India, Bangladesh and Pakistan.

In these countries, the dismantling of ships that are out of service is performed without any protection to the environment and the people involved, resulting in a high rate of mortality and chronic diseases.

The workers of the ship-breaking yards are being exploited, abusing their basic necessities in life, and desperation at the lack of work. This labour is cheap and abundant in these facilities, compared with the price to pay to the owner of a ship to be scrapped, which is much higher in other countries, and represents very stiff competition [5].

The governments of the countries that should protect these workers and their beaches, dictating rules on occupational hazards and protection of the environment, however, was when India began to ask for a free certificate to tankers arriving to be scrapped. This is when it became clear that this action should be jointly between the maritime world and the government of the respective country, when there was this initiative by India, Pakistan and Bangladesh. These countries did not support it, which resulted in a movement of tankers that were to be scrapped to these countries.

If this action is not performed simultaneously to try to improve safety conditions in the yards of these countries, there could be another side effect to ship dismantling, by way of a shift towards countries that are today less developed countries [4].

The scrap yards in Asia are benefiting also by how expensive it is to promote a recycling plant in a developed country, which requires permits and environmental impact studies to be viable. With this investment, developed countries could never pay for a ship to be recycled at the prices currently being paid.

This is why the OECD countries lack the capacity to dismantle vessels of the member states of the EU. However, there are countries outside the OECD with safe recycling capacities to handle vessels of the member states. This conflicts directly with the Basel Convention, which, considering the ship as a hazardous waste, prohibits their movement from one OECD country to another that is not a member [5].

In these situations so contradictory:

- Need for dismantling old ships.
- Lack of dismantling facilities in the OECD.
- Prohibition of moving toxic waste from an OECD country to another that is not a member.

That is, the vessel needs to be scrapped, but the OECD does not have capacity to do so, but neither can it be moved outside the OECD. This, coupled with the large price paid by a ship dismantling unit in Asian countries, has made the Basel Convention appear to be continuously failing. The way to bypass this Convention is to change the flag of a vessel of a member state of the EU, to one that does not belong to the EU, or to sell the ship to a buyer who will sign a declaration stating that the ship is not intended for dismantling, but when it comes out of European waters will head to India, Pakistan or Bangladesh, ignoring this declaration [3]. With the new Agreement, the Hong Kong Convention will allow these countries, once it has been signed, and their dismantling facilities have been improved, to be accepted for recycling without contradicting the Basel Convention, preserving its main working environment [5]. This Convention establishes, first, the preparation of ships for a safe process, and that which respects the environment, and on the other hand, establishes the operation of ship dismantling facilities [17]. The aim is that countries that ratify scrapping of their vessels, provided they have a certificate of suitability for recycling, in facilities approved by a member state. It would be necessary, including in Asian countries that signed the Hong Kong Convention, to ensure that its facilities meet safety standards, treatment records that carry toxic substances, as these countries do not belong to the Basel Convention. In addition, the owners of ships being dismantled should be aware that the price of these vessels will fall if the improvements that are intended to be achieved to avoid deaths continue to occur, and accidents keep taking place in their yards. Finally, since June 2007, the EU has a financial instrument called Life+, which aims to support projects that improve the development, updating and implementation of community policies and regulations that deal with the environment. Life+ funded a project called RECYSHIP for dismantling and decontamination of ships taken out of use. The European Commission intends, with this project, to end the controversy on ship recycling in Europe and solve the problem of the transfer of ships to countries like India, Pakistan or Bangladesh. Their objectives include:

- Developing a method to make the scrapping of vessels a safe activity, environmentally friendly and profitable in Europe.

- Creating prototypes to decontaminate and scrap service vessels which are not to be used later on in other ships.
- Checking the capacity and impact of these activities in southern Europe.
- Assisting in the development of legal measures to take care of the waste.
- Encouraging the use of sustainable technologies in the countries where they undertake scrapping [4].

To achieve this is required a pilot plant consisting of:

- A pilot plant extraction of hazardous waste.
- An automated pilot plant for extraction of fluid pollutants: hydrocarbons (oil and fuel), bilge water, ballast and other fluids.
- A pilot plant extraction of TBT.
- A pilot plant for non-hazardous waste removal [5].

In June 2012, Recyship made the decontamination of a ship taken out of use, in Portugal. To do this, they used a "tick robot" designed to perform caesarean sections in the ship's hull and engines for extraction, and so on.

They have also designed a paint-stripping machine that cleans the steel sustainably, and have a treatment system so as to effectively separate water and oils [18].

REFERENCES

[1] Rodrigo de Larrucea, Jaime 2009, El convenio internacional de Hong Kong para la seguridad y gestión medioambientalmente racional del reciclado de buques Hong Kong.

[2] Commission Staff Working Document - Accompanying document to the Green Paper on better ship dismantling, {COM(2007) 269 final}.

[3] Siecker, Martin 2012, Proposal for a Regulation of the European Parliament and of the Council on ship recycling, COM(2012) 118 final – 2012/0012 (COD), *Plenary Session*: 482 - 11 Jul 2012 - 12 Jul 2012.

[4] Work group, 2008, Sostenibilidad de los desguaces de buques (GT-BUQ), 9th *Congreso Nacional del Medio Ambiente.*

[5] Núñez Basáñez, J.F., Gutiérrez Fraile, R. 2012, "Una industria sostenible de desguace y reciclado de buques", *XXVIII Semana de estudios del Mar*, pp. 233-263.

[6] Dictionary of Nautical Wordsand Terms, by C.W.T. Layton, F.R.A.S., M.R.I.N., Assoc.R.I.N.A., Glasgow Brown, SonandFerguson, LTD., Nautical Publishers.

[7] Vistaalmar. http://www.vistaalmar.es/medio-ambiente/contaminacion /2418-dentro-astilleros-desguace-barcos-chittagong.html (Accesed November 2012).

[8] Imfmetal. http://www.imfmetal.org/index.cfm?c=28154andl=28 (Accesed November 2012).

[9] Industriall-union. http://www.industriall-union.org/es/una-vez-mas-mueren-trabajadores-en-un-astillero-de-desguace-de-buques-de-la-india (Accesed November 2012).

[10] Basel Convention, on the control of transboundary movements of hazardous wastes and their disposal, www.basel.int, Secretariat of the Basel Convention.

[11] Senado español, X Legislatura, Registro general entrada 2.666 15/02/2012.

[12] Summary of the Tenth Meeting of the Conference of the parties to the Basel Convention, October 2011, vol. 20, no. 3, pp. 1-16.

[13] Mikelis, Nikos 2010, "Introduction to the Honk Kong Convention and its requirements", *Ship Recycling Technology and Knowledge Transfer Workshop Organized by the Secretariat of the Basel Convention*, International Maritime Organization.

[14] Boletín Informativo de ANAVE Nº487 (Gabinete de estudios de ANAVE) 2009, "Adoptado en Hong Kong el nuevo convenio de la OMI sobre reciclaje de buques".

[15] Mikelis, Nikos 2010, "The Hong Kong International Convention for the Safe and environmentally sound recycling of ships", *United Nations Conference on trade and development*, Multi-year expert meeting on transport and trade facilitation.

[16] NGO Shipbreaking Plataform September 2011, "Statemet of concern to the parties of the Basel Convention on the Hong Kong Convention on ship recycling".

[17] Proposal for a "Regulation of the European Parliament and of the council on ship recycling", {COM(2012) 118 final}.

[18] http://www.europapress.es/sociedad/medio-ambiente-00647/noticia-reciclauto-navarra-inicia-trabajos-descontaminacion-barcos-portugal-20120730202337.html (Accesed November 2012).

In: Ships and Shipbuilding
Editor: José A. Orosa

ISBN: 978-1-62618-787-0
© 2013 Nova Science Publishers, Inc.

Chapter 8

IMO STANDARD AND GAS EMISSIONS REDUCTION FROM SHIPS

Ángel M. Costa,[*] Rebeca Bouzón, A. De Miguel Catoira, J. Romero Gómez and M. Romero Gómez
Department of Energy and Marine Propulsion
University of A Coruña, Spain

ABSTRACT

Nowadays, maritime transport shows an increase in its restrictions about gases emission into the atmosphere. These restrictions are given by the International Maritime Organization through the Annex VI of MARPOL. It is due to ship engines produce different gases which are related with greenhouse effect and causing different pollution types.

The propulsion energy employed in sea transport is obtained today, nearly exclusively, by fossil fuels derived from petroleum whose combustion produces emissions such as CO_2, NO_X and SO_X.

In consequence, although shipping has the lowest ratio of CO_2 emissions per ton carried per mile than other modes of transport, it is expected these emissions of greenhouse effect gases will increase in coming decades.

Over 90% of world trade is carried on board ships and, although it is a high efficient transport way, it should be noted the large quantity of energy that is used for such purposes, and using this energy will produce exhaust gases with all its effects on the environment and quality of life, therefore these areas of research are important.

Besides the impact of CO_2 in the atmosphere, it is also considered to reduce NO_X, SO_2 and other harmful gases emissions for propulsion systems and human beings placed at sea or in port areas and its global environmental impact. This review outlines that articles considered more important in recent years related to this attempt to reduce the ships' emissions.

---

[*] Email: jaorosa@udc.es

INTRODUCTION

The International Convention for the Prevention of Pollution from Ships, or MARPOL 73/78, is a set of international standards laid out in order to reduce and/or eliminate pollution emanating from ships. It was developed by the International Maritime Organization (IMO) which is a specialised agency of the United Nations (UN).

Marine pollution 1973–1978 (MARPOL 73/78), was originally adopted in 1973, but was never enforced. The main matrix of the current version was modified by the Protocol of 1978, and since that time, it has been modified by a great number of corrections. MARPOL finally entered into enforcement on October 2, 1983, and currently, 119 countries have ratified it. It aims to preserve the marine environment through the total elimination of pollution by oil and other harmful substances from sea-going vessels, and the minimisation of potential accidental discharges caused by them.

If we now analyse the MARPOL Convention structure, we can see that the Agreement consists of an introduction, the inherent text of the International Convention for the Prevention of Pollution from Ships (1973), the protocol of 1978 related to that of 1973, the Protocol of 1997 to amend the 1973 Convention as amended by the 1978 Protocol, and six annexures consisting of rules covering the various sources of pollution from ships, as shown below.

- Annex I: Regulations for the Prevention of Pollution by Oil.
- Annex II: Regulations for the Prevention of Pollution by Noxious Liquid Substances in Bulk.
- Annex III: Regulations for the Prevention of Pollution by Harmful Substances Carried by Sea in Packages. This is optional, in that the transport of dangerous goods is regulated by the International Maritime Dangerous Goods Code.
- Annex IV: Regulations for the Prevention of Pollution by Sewage from Ships.
- Annex V: Regulations for the Prevention of Pollution by Garbage from Ships.
- Annex VI: Regulations for the Prevention of Air Pollution from Ships. This Annex entered into force on May 19, 2005, and will be analysed in the next section.

1. ANNEX VI

Although land-based emissions in Europe are on the decline, emissions from ships continue to grow [1], which can spread to the rest of the world. In addition to this, ship emissions cause air quality problems in coastal areas and harbours that see heavy sea traffic [2]. Annex VI of MARPOL was constituted mainly to control these emissions. This Annex has joined the Convention by the Protocol of 1997, which was signed in London on September 26, 1997, remaining open to ratification by the various parties on January 1, 1998. The Rules of Annex VI are grouped into several chapters that cover the following topics:

- Chapter I. General (Rules: 1–4).
- Chapter II. Recognition, certification and control means (Rules 5–11)
- Chapter III. Requirements for the control of emissions from ships (Rules 12–19).

- This also includes five appendices dealing with:
- Appendix I. Model IAPP Certificate.
- Appendix II. Test cycles and weighting factors.
- Appendix III. Criteria and procedures for the designation of areas of SOx emission control.
- Appendix IV. Approval and service limits for shipboard incinerators.
- Appendix V. Information to be included in the bunker delivery note.

2. OBJECTIVE

Emission of exhaust gases and particles from ships are a significant and growing contributor to the total emissions from the transportation sector [3]. This chapter presents an assessment of the contribution of gaseous and particulate emissions from ships to anthropogenic emissions and air quality. It is also revised the bibliography about the degradation brought about in human health and climate change as a direct result of these emissions. Thus, the regulation of ship emissions requires comprehensive knowledge of current fuel type, consumption and emissions, understanding of their impact on atmospheric composition, climate and projections of potential future evolutions and mitigation options.

As we can see, a reliable and up-to-date ship emission inventory is essential to assist atmospheric scientists in quantifying the impact of shipping for policy makers to implement necessary regulations and incentives for emission reduction [4].

Hence, the existing and emerging international and European policy framework for the reduction of ship exhaust emissions dictates the need to produce reliable national, regional and global inventories in order to monitor emission trends and, consequently, provide the necessary support for future policy making [5].

3. MATERIALS AND METHODS

The methodology to achieve the previously mentioned objective consists of a classification of scientific articles based on the different rules contained in MARPOL Annex IV. These rules will be described now. Rule 1 is applied to all ships, except those ships that are cited in the other rules. In Rule 2, new terms that are not listed in the rules of the other Annexes are introduced, such as power supply, emissions, NOx Technical Code, substances that deplete the ozone layer, emission control area, SOx, and so on. In particular, Rule 12, titled "Substances that Deplete the Ozone Layer" prohibits deliberate emissions of substances that deplete the ozone layer, including emissions during maintenance, servicing, repair or decommissioning of equipment and systems that contain them. Exceptions are when releasing quantities of material for recycling, or recovery of a substance. Substances that deplete the ozone layer must not be used in new installations on board, except in the case of hydrochlorofluorocarbons (HCFCs), which are permitted until January 1, 2020. These substances, and the equipment containing them shall be kept in appropriate retention facilities when they are removed from ships. The tool in the research work [6] is a global atmospheric chemistry transport model which simulates the chemical state of the Mediterranean atmosphere when applying different ship emission inventories. Close to the major shipping routes, relative contributions vary from 10% to 50% for black carbon, and from 2% to 12% for ozone in the surface layer, as well as from 5% to 20% for nitrogen dioxide atmospheric column burden. Measurements of aerosol chemical compositions made in the south of the Sicily channel are used to identify the influence of ship emissions on aerosol particles in the Central Mediterranean. Evidence of the influence of ship emissions is found in 17% of the daily samples [7]. The influence that ship emissions exert on ozone (O3) concentrations in a coastal area was examined based on a numerical modelling approach during a high O3 episode. The model study [8] suggested the possibility that pollutant gases emitted from ships can have a direct impact on the O3 concentration levels in the coastal area. The greatest impacts of ship emissions on the O3 concentrations were predicted at the coast (up to 15 ppb) and at inland locations (about 5 ppb) due to both the photochemical production of pollutant gases emitted from the ships and meteorological conditions.

3.1. Rule 13. Nitrogen Oxides (NOx)

This rule must be applied to:

- Each diesel engine with output power greater than 130 kW, installed on a ship constructed on January 1, 2000 or later.
- Each diesel engine with output power greater than 130 kW, which has undergone a major conversion on or after January 1, 2000.

The rule must not be applied to:

- The emergency diesel engines, to those installed in lifeboats and, in general, to all devices or equipment designed to be used only in emergencies.
- The engines installed on ships engaged exclusively for traffic within waters under the jurisdiction or sovereignty of the flag state, as long as these engines are subject to measurements of NOx control established by the administration.

The same rule determines that the most important transformation in the engine involves:

- The replacement by a new one built on January 1, 2000 or later.
- The completion of a substantial change, as defined in the NOx Technical Code.
- Increasing the maximum continuous speed by over 10%.

The rule also prohibits the operation of a diesel engine unless it meets certain emission limits for nitrogen oxides calculated using mathematical equations that are included in the same section. Emission measurements were made for major gases and (particulate matter) PM 2.5 mass for a post-PanaMax Class container vessel operating on heavy fuel oil at sea. Results showed the weighted emission factor for NOx and PM2.5 were 19.77 ± 0.28 and 2.40 ± 0.05 g/kW/h, respectively. Emission factors of CO2 and NOx closely matched the earlier values, suggesting a low deterioration factor [9]. The NOx reductions from the exhaust [10] can be as a consequence of fuel nitrogen content and engine operation; the PM2.5 reduction is attributed to the large reductions in the hydrated sulphate and organic carbon fractions. As expected, operating in the low NOx mode reduced NOx emissions by approximately 32%, and nearly doubled the elemental carbon emissions. In the investigation [11], the emission inventories for 2000 and 2015 are used in a global Chemical Transport Model to quantify environmental atmospheric impacts with particular focus on the Arctic region. Although we assume that ship emissions continue to increase from 2000 to 2015, reductions are assumed for some chemical components and regions because of implementation of the new regulations. Therefore, current ship traffic is estimated to contribute significantly to coastal pollution. For example, Norwegian coastal ship traffic is responsible for more than 1/3 and 1/6 of the Norwegian NOx and SO2 emissions, respectively. At the same time, the brief lifetime of the NOx plume over the entire ship clearly shows that the ship-plume chemistry shortens the NOx lifetime considerably. Therefore, the ship-plume chemistry model should be used to model the changes in ship-plume chemical compositions and better evaluate the atmospheric impact of ocean-going ship emissions [12]. Thus, observations have demonstrated a NOx lifetime to be 1.8 hours inside the ship's plume, compared with 6.5 hours (at noontime) in the moderately polluted background marine boundary layer of the experiment. This confirms the hypothesis [13] of highly enhanced in-plume NOx destruction. Consequently, one would expect that the impact of ship emissions is much less severe than those predicted by global models that do not include rapid NOx destruction. Ship emissions are significantly increasing globally, and have made a remarkable impact on air quality on sea and land. In particular, these emissions contribute adversely against health and the environment. Parameters, like main engine, fuel, operations, navigation times and speeds of the vessels are taken into consideration in the study [14] as the main parameters with which to analyse ship emissions. Results showed that shipping emissions in the Sea of Marmara and the Turkish Straits are equivalent to 11% of NOx, 0.1% of CO, and 0.12% of PM of the corresponding total emissions in Turkey. Finally, more research works showed that the impact of ship emissions on tropospheric oxidants is mainly caused by the relatively large fraction of NOx in ship exhausts, and typical increases in the yearly average surface ozone concentrations in the most impacted areas, which are 0.5–2.5 ppb. Furthermore, the findings in other related research works [15] support the earlier observational studies indicating that ship traffic may be a major contributor to recent enhancement of background ozone at some coastal stations.

3.2. Rule 14. Sulphur Oxides (SOx)

Rule 14 establishes the general requirements applicable to all vessels, that the fuel oil used as fuel must contain no more than 4.5% sulphur. In the case of so-called Zones SOX emission control, this provides some special requirements. The conditions for ships operating within a control zone are:

- The fuel SO_x content should not exceed 1.5%.
- If using a gas cleaning system approved by the administration to reduce emissions of sulphur oxide, its content should not exceed 6 g of SO_x/kW/h, and waste streams

should not be discharged into enclosed ports or estuaries, unless it is proved, with valid documents, that these are not harmful to the ecosystem of the area.

- If you use any other method of reduction, it must be approved by the administration, and the level of sulphur emissions should be similar to the previous one.

Particulate matter emissions from one serial 4-stroke medium-speed marine diesel engine were measured for load conditions from 10% to 110% in test rig studies using heavy fuel oil [16]. Testing the engine across its entire load range permitted the scaling of exhaust particulate matter properties with load. The engine load dependence of the conversion efficiency for fuel sulphur into sulphate of 1.08% ± 0.15% at engine idling, to 3.85% ± 0.41% at cruising speed, may serve as input to global emission calculations for various load conditions.

The model [17] shiws the cost-effectiveness of control strategies for reducing SO2 emissions from US foreign commercial ships travelling in existing European or hypothetical US West Coast SOx Emission Control Areas (SECAs) under international maritime regulations. Variation among marginal costs of control for individual ships choosing between fuel-switching and after treatment reveals cost-saving potential of economic incentive instruments.

The Article [18] estimates that the 172,000 ship voyages to and from North American ports in 2002 consumed about 47 million metric tonnes of heavy fuel oil, and emitted about 2.4 million metric tonnes of SO2.

Significant variations are apparent between the various reported regional and global ship SO2 emission inventories. Important parameters for SO2 emission modelling are sulphur content and marine fuel consumption. The paper [19] presents an improved bottom-up approach to estimate marine sulphur emissions from ship transportation, including the geographical distribution. The Article presents an alternative approach to estimate present overall SO2 ship emissions with much improved accuracy.

The study [20] investigates the impact of ship emissions in coastal areas of the North Sea. The implementation of a SECA in the North Sea, as it was implemented at the end of 2007, directly results in reduced SO2 and sulphate aerosol concentrations, while nitrate aerosol concentrations are slightly increased.

Atmospheric emissions of gas and particulate matter from a large ocean-going container vessel were sampled as it slowed and switched from high-sulphur to low-sulphur fuel when the vessel transited into the regulated coastal waters of California. Reduction in emission factors of SO2, particulate matter, particulate sulphate and cloud condensation nuclei were substantial (≥90%).

The analysis presented [21] provides direct estimations of the reductions in emissions that can be realised by California fuel quality regulation and the VSR program, in addition to providing new information relevant to potential health and climate impact of reduced sulphur content of the fuel, fuel quality and vessel speed reductions.

Potential costs and benefits of policy options for reducing offshore ship pollution are examined using a meta-analysis of studies synthesised regionally for the US West Coast [22]. The results show that about 21,000 tons of on-land-equivalent SO2 emissions, or about 33% of SO2 emissions from all mobile sources in California, can be reduced annually if the US West Coast exclusive economic zone is designated as an International Maritime Organization-compliant SECA with fuel sulphur content not exceeding 1.5%. Technological

IMO Standard and Gas Emissions Reduction from Ships

alternatives may achieve benefits of emissions reductions with higher ranges of potential fuel prices. Combinations of fuel-switching, and control technology strategies provide the most cost-effective benefits from SECAs.

3.3. Rule 15. Volatile Organic Compounds

This refers to the possible regulation by the port state, emissions of volatile organic compounds (VOCs) from tankers, and determines that:

- The regulations must be in accordance with this rule.
- States' Parties intending to regulate the emission of VOCs are required to notify the IMO on vessel size and controlled loads that require it, and the date of entry into force of the measure.
- The Governments of the Parties to the Protocol of 1997 to designate ports or terminals where it is implemented to regulate VOC emissions, shall ensure that such places exist in control systems approved by the administration in accordance with the rules relating to security developed by IMO (MSC/Circ.585).
- IMO distributed to all other Member States a list of ports designated by each Party to control VOC emissions.
- Tankers will be monitored, and must be equipped with a vapour collection system approved by the Administration in accordance with the safety standards developed by the IMO. In cases where the operations are conducted in terminals equipped with a control system, existing vessels that do not have a vapour collection system for a period of three years after the entry into force of that control may be accepted.

Particle emissions from ship engines and their atmospheric transformation in the marine boundary layer were investigated in engine test-bed studies, and in airborne measurements of expanding ship plumes.

During the test-bed studies [23], detailed aerosol microphysical and chemical properties were measured in the exhaust gas of seven-cylinder four-stroke marine diesel engines under various load conditions. The emission studies were complemented by airborne aerosol transformation studies in the plume.

Particulate matter emission indices obtained from plume measurements are $8.8 \pm 1.0 \times 10$ 15 (kg fuel) -1 by number for non-volatile particles, and 174 ± 43 mg (kg fuel) -1 by mass for black carbon.

3.4. Rule 16. Shipboard Incineration

Incineration is allowed only in one incinerator. Incinerators installed on or after January 1, 2000, shall comply with the requirements of Appendix IV of this Annex, and shall be approved by the administration, taking into account the standard specifications for shipboard incinerators developed by the IMO (resolution MEPC.76 (40)).

The administration may waive application of regulations on the previous paragraph to incinerators installed before the date of entry into force of the Protocol of 1997, provided that the ship is engaged only on voyages within waters over which the flag state exercises sovereignty or jurisdiction.

It prohibits the on-board incineration of the following substances:

- Waste loads listed in Annexes I, II, and III of the Convention, and their contaminated packaging.
- Polychlorinated biphenyls (PCS).
- The wastes containing heavy metals at concentrations greater than mere traces.
- Refined petroleum products containing halogen compounds.

It prohibits the on-board incineration of polyvinylchloride (PVC), except in incinerators that have a certificate of approval from the IMO.

No incineration is to be below 850°C, and in the batch-load incinerators, the combustion chamber must reach 600°C in the first 5 minutes of ignition.

3.5. Rule 17. Reception Facilities

The states' Parties to the 1997 Protocol undertake to provide its port facilities to be free of substances that deplete the ozone layer, and the equipment that contains them, and the appropriate means for scrapping when this is taken on board. Furthermore, facilities should be provided for receiving waste from the cleaning of exhaust gases when vessels use a port or terminal, or port repairs require it.

3.6. Rule 18. Quality Fuel Oil

Except for solid coal and nuclear fuels, fuel oil used on board must comply with the conditions set forth in this Annex concerning the sulphur content, and that cause the engine to exceed the limits of NOx determined in Rule 13. Neither should it contain organic acids or chemical wastes or substances that are harmful to:

- The safety of the ship
- The performance of the engines
- Persons on board, or
- Contribute to additional air pollution.

Fuel suppliers are obliged to provide the vessel a delivery note containing at least the information set out in Appendix V of this Annex.

All these investigations contribute to an improved understanding of the impacts of ship emissions on climate and air quality, and will also assist in determining potential effects of altering the present fuel standards [24]. In particular, composition of exhaust gasses from a ship diesel engine using heavy fuel oil was investigated on board a large cargo vessel [25]. In this investigation, the PM composition was dominated by organic carbon, ash and sulphate, while the elemental carbon composed only a few percent of the total PM. Increase of the PM in the exhaust upon cooling was associated with increase of organic carbon and sulphate. Laser analysis of the adsorbed phase in the cooled exhaust showed presence of a rich mixture of polycyclic aromatic hydrocarbon species with molecular mass 178–300 amu, while PM collected in the hot exhaust showed only 4 polycyclic aromatic hydrocarbons (PAH) masses.

Other research works [26] explained the 2006–2020 emission forecasts, and demonstrated a need for stricter fuel quality and NOx emission standards for navigation, in order to gain emission improvements in line with those achieved for other mobile sources.

More research works [27] provided the emission factors for selected PAH, heavy alkanes, carbonyls, light hydrocarbon species, metals and ions for the main engine, auxiliary engine and the auxiliary boiler. A deeper analysis showed that the conversion of fuel sulphur to sulphate in the main engine ranged from 1.4% to 5%.

Composition of exhaust from a large cargo vessel was investigated on board a ship during the Quantify field campaign [28]. Hazardous constituents from the combustion of heavy fuel oil, such as transitional and alkali earth metals (V, Ni, Ca, Fe), were observed in the particulate matter samples.

Extensive measurements [29] of the emission of light-absorbing carbon aerosol from commercial shipping are presented. The highest emitters (per unit fuel burnt) are tug boats, thus making significant contributions to local air quality in ports. Emission of light-absorbing carbon aerosols from cargo and non-cargo vessels in this study appears to be independent of engine load, by what may be due to the quality of fuel used.

The work [30] presents the results of field emission measurements that have been carried out on the 4500 kW four-stroke main engine on board a fuel product tanker. Two fuel qualities (heavy fuel oil and marine gas oil) have been tested on the same engine for comparable load settings. Specific PM emissions were generally higher for heavy fuel oil than for marine gas oil; however, for the smallest size of fraction measured containing particles 0.30–0.40 µm in diameter, the opposite is observed. This finding emphasises that to minimise negative health effects of exhaust particles from ships, further regulation may be needed to reduce small-sized particles; a fuel shift to low-sulphur fuel alone does not seem to accomplish this reduction.

Key physico-chemical characteristics of diesel exhaust particulates of sea-going ship emissions are presented with respect to morphology, microstructure, and chemical composition. Heavy fuel oil-derived particles exhibit extremely complex chemistry. The composition analysis investigates the content of environmentally dangerous pollutants: metals, inorganic/mineral species and soluble, volatile organic and ionic compounds. It is found [31] that hazardous constituents from heavy fuel oil combustion, such as transitional and alkali earth metals (V, Ni, Ca, Fe) and their soluble or insoluble chemical forms (sulphides, sulphates, oxides, carbides), are released together with particles into the atmosphere.

Single particles containing mixtures of organic carbon, vanadium and sulphate (OC-V-sulphate) resulted [32] from residual fuel combustion (i.e., bunker fuel), whereas, high quantities of fresh soot particles (when OC-V-sulphate particles were not present) represented distinct markers for plumes from distillate fuel combustion (i.e., diesel fuel) from ships as well as trucks in the port area. OC-V-sulphate particles from residual fuel combustion contained significantly higher levels of sulphate and sulphuric acid than plume particles containing no vanadium.

3.7. Other Issues not Listed in Annex VI

Greenhouse gas emissions from international maritime transport are exempt from liabilities under the Kyoto Protocol and MARPOL. Research into quantifying these emissions is ongoing and influences policy proposals to reduce emissions.

International shipping is a significant contributor to Global Greenhouse Gas emissions, responsible for approximately 3% of global CO2 emissions. The IMO is currently working to establish greenhouse gas regulations for international shipping, and a cost-effective approach has been suggested to determine the required reductions in emissions from shipping.

The work [33] presents a cargo-based analysis of fuel consumption and greenhouse gas emissions. A cargo-based methodology was used to estimate that the international maritime transport of New Zealand's imports and exports consumed 2.5 million tonnes of fuel during the year 2007, which generated 7.7 million tonnes of carbon dioxide (CO_2) emissions.

The method used in the article [34] is an extractive sampling method of the passing ship's plumes, where particle number/mass and CO2 were measured with high time resolution. To achieve reductions in emissions in a cost-effective manner, the study [35] has assessed the cost and reduction potential for present and future abatement measures based on new and unpublished data. The model used captures the world fleet up 2030, and the analysis includes 25 separate measures. A new integrated modelling approach has been used combining fleet projections with activity-based CO2 emission modelling and projected development of measures for CO_2 emission reduction.

Thus, the paper [36] investigates the economic costs of CO_2 emissions reduction with and without considering the CBDR under two separate regulatory scenarios, evaluates the impact on trade costs, and discusses the real policy concerns of developing countries insisting that the CBDR must be applied by the IMO.

In the paper [37], the various options suggested by the Subsidiary Body for Scientific and Technological Advice of the UNFCCC for allocating CO_2 emissions from international shipping to individual countries are investigated. This is followed by a discussion of the economic and regulatory issues related to these options, and the consequences of applying them.

On the other hand, different contamination sources must be analysed. In this sense, in the research work [38], three likely sources of the elevated HCHO levels in ship plumes, as well as their contributions to the elevated HCHO levels (budget), were investigated using a newly developed ship-plume photochemical/dynamic model.

4. REDUCING EMISSIONS

Despite the increase in commercial shipping around the world, data are still relatively scarce on the contribution of these emissions to ambient air particulates. One of the reasons [39] is the complexity in the detection and estimation of shipping contributions to ambient particulates in harbour and urban environments, given the similarity with tracers of other combustion sources.

The emissions of CO_2, SO_2, NO_2 and PM from maritime transportation have contributed to climate change and environmental degradation. Scientifically, analysts still have controversies regarding how to calculate the emissions and how to choose the baseline and methodologies. This, in turn, results in great difficulties to policy makers who attempt to regulate the emissions. However, the regulations based on its results may increase the costs of shipping companies and result in competitiveness of the port states and coastal states. The paper [40] concludes with policy indications in the future, and work that needs to be done.

In the study [41], the size and composition of freshly emitted individual ship exhaust particles has been investigated using an aerosol time-of-flight mass spectrometer collocated with a suite of real-time instrumentation. The results suggest that aerosol time-of-flight mass spectrometer single particle mass spectra, when used in conjunction with other air quality monitoring instrumentation, may be useful in determining the contribution of local shipping traffic to air quality in port cities.

The paper [42] provides a critical analysis of the ship emission modelling approaches and data sources available, identifying their limits and constraints. As a final remark, it can be expected that new approaches to the problem, together with more reliable data sources over the coming years, could give more impetus to the debate on the global impact of maritime traffic on the environment that, currently, has only reached agreement via the "consensus" estimates provided by IMO.

The paper [43] summarises first, the results of the impact of ship emissions like NO_2 and aerosol on climate and chemistry, as determined from satellite data. The results highlight the importance of ship emissions for the marine boundary layer, and at the same time, demonstrates the potential of satellite observations to estimate the global impact.

The study [44] concludes that the main reason for the large deviations found in reported inventories is the applied number of days at sea. The model indicates that the size of the ship, and the degree of utilisation of the fleet, combined with the shift to diesel engines, have been the major factors determining yearly fuel consumption. Interestingly, the model results from around 1973 suggest that fleet growth is not necessarily followed by increased fuel consumption, as technical and operational characteristics have changed. Results from this study indicate that reported sales over the last three decades seems not to be significantly under-reported, as previous simplified activity-based studies have suggested.

The dispersion and chemical conversion of emissions in the near-field of a single ship are studied with two different modelling [45] approaches to explore the differences between gradual dispersion and instantaneous dilution into a box with a size comparable to the large grid boxes of global-scale models or satellite data. One approach uses a Gaussian plume model, and accounts for the expansion phase of a plume. The other one instantaneously disperses the emissions over a large grid box, a technique commonly used by large-scale models.

It is also important to evaluate the cost and effectiveness of improvements in emissions reduction technologies on ships. This is also necessary to analyse the emerging scientific evidence for all decisions in the management of environmental issues related to this area.

The article [46] evaluates six black carbon emissions reduction technologies for marine engines, including the net effect on a set of short-lived climate forces from marine diesel combustion. Technologies evaluated include slide valves, water-in-fuel emulsion, diesel particulate filters, low-sulphur fuel, emulsified fuel and sea water scrubbing.

The article [47] presents a global bottom-up ship emission algorithm that calculates fuel consumption, emissions and vessel traffic densities. The developed algorithm automatically finds the most probable shipping route for each combination of origin and destination port of a certain ship's movement by calculating the shortest path on a predefined model grid, while considering land masses, sea ice, shipping canal sizes, and climatological mean wave heights.

A method is presented in the paper [48] for the evaluation of the exhaust emissions of marine traffic, based on the messages provided by the Automatic Identification System (AIS), which enable the identification and determining the locations of ships. The use of the AIS data facilitates the positioning of ship emissions with a high spatial resolution. The modelling system can be used as a decision support tool in the case of issues concerning, e.g., the health effects caused by shipping emissions, or the construction of emission-based fairway dues systems or emissions trading.

Ship activity patterns and their combination, demonstrate different spatial and statistical sampling biases. These differences could significantly affect the accuracy of ship emissions inventories and atmospheric modelling. The paper [49] demonstrates a method to improve global-proxy representativeness by trimming over reporting vessels that mitigates sampling bias, augments the sample dataset, and accounts for ship heterogeneity.

In the study [50], the size distribution of particles in ship exhaust from three different ships in normal operational conditions were studied using a cascade impactor. The particle emissions were found to be reduced to about half, over the whole size range, by an SCR system. The total particle emission, measured after dilution, varied between 0.3 and 3 g kW/h depending on load, fuel and engine.

Recognising that associating impacts with black carbon emissions requires both ambient and on-board observations, we provide recommendations for the measurement of black carbon. The paper [51] also evaluates current insights regarding the effect of ship speed (engine load), fuel quality and exhaust gas scrubbing on black carbon emissions from ships. Uncertainties among current observations demonstrate there is a need for more information on (a) the impact of fuel quality on black carbon emission factors using robust measurement methods, and (b) the efficacy of scrubbers for the removal of particulate matter by size and composition. The potential of a load-levelling strategy through use of a hybrid battery-diesel-electric propulsion system is investigated [52]. The goal is to reduce exhaust gas emissions by reducing fuel oil consumption through consideration of a re-engineered ship propulsion system.

Finally, the paper [53] examines the role of marine engine maintenance in reducing pollution. This paper explains how the condition of an engine's nozzles and faulty injection pressure significantly influence NOx and CO emissions, and describes both bench and onboard ship tests, on engines fitted with new or worn nozzles at different injection pressures.

CONCLUSION

A steady increase in the number of articles published on air pollution contained in Annex IV of MARPOL has been observed in the last 12 years. From this review, it can be concluded that the emissions control on ships is becoming more and more important, as they approach the deadline to fully enter into force. Although emissions control in ground facilities is much easier than on ships, monitoring of these emissions are equally important in more ground facilities.

The goal of an almost complete suppression of the sulphur content in fuels used in shipping is the decision of the IMO of the UN, to drastically reduce the sulphur content of marine fuels by 2020. Therefore, everyone should rush to the members of this organisation to ratify the Convention, thus achieving a global application.

Further research is needed in improving alternative methods of reducing emissions or efficiency of energy generation from fossil fuels, or switching to alternative fuels, taking into account the various uncertainties, such as the availability of fuel with low sulphur content, in 2015.

It should also continue research on the impact of air pollution on maritime traffic in SECAs, and see its real impact on health of marine and terrestrial habitats, as also in other areas, although not so delicate, but equally important.

REFERENCES

[1] Devasthale, A., Krüger, O. and Graßl, H., (2006). Impact of ship emissions on cloud properties over coastal areas. *Geophysical Research Letters*, 33, 2.

[2] Eyring, V., (2008). Modelling of the coupled chemistry-climate system: Projections of stratospheric ozone in the 21st century and impact of shipping on atmospheric composition and climate. http://www.pa.op.dlr.de/~VeronikaEyring/Habilitation_ VEyring_Forschungsbericht_FINAL.pdf (accessed december 2012).

[3] Eyring, V., Isaksen, I.S.A., Berntsen, T., Collins, W.J., Corbett, J.J., Endresen, O., Grainger, R.G., Moldanova, J., Schlager, H. and Stevenson, D.S. (2010). Transport impacts on atmosphere and climate: Shipping. *Atmospheric Environment*, 44, 37, 4735-4771.

[4] Dalsøren, S.B., Eide, M.S., Endresen, O., Mjelde, A., Gravir, G. and Isaksen, I.S.A. (2009). Update on emissions and environmental impacts from the international fleet of ships: The contribution from major ship types and ports, *Atmospheric Chemistry and Physics*, 9, 6, 2171-2194.

[5] Tzannatos Ernestos, E., (2010). Ship emissions and their externalities for Greece, *Atmospheric Environment*, 44, 18, 2194-2202.

[6] Manner, E., Dentener, F., V Aardenne, J., Cavalli, F., Vignati, E., Velchev, K., Hjorth, J., Boersma, F., Vinken, G., Mihalopoulos, N. and Raes, F. (2009). What can we learn about ship emission inventories from measurements of air pollutants over the Mediterranean Sea?, *Atmospheric Chemistry and Physics*, 9, 18, 6815-6831.

[7] Becagli, S., Sferlazzo, D.M., Pace, G., Di Sarra, A., Bommarito, C., Calzolai, G., Ghedini, C., Lucarelli, F., Meloni, D., Monteleone, F., Severi, M., Traversi, R. and Udisti, R. (2011). Evidence for ships emissions in the Central Mediterranean Sea from aerosol chemical analyses at the island of Lampedusa, *Atmospheric Chemistry and Physics Discussions*, 11, 11, 29915-29947.

[8] Song, S., Shon, Z., Kim, Y., Kang, Y., Oh, I.. and Jung, C. (2010). Influence of ship emissions on ozone concentrations around coastal areas during summer season, *Atmospheric Environment*, vol. 44, 5, 713-723.

[9] Agrawal, H., Welch, W.A., Henningsen, S., Miller, J.W. and Cocker III, D.R. (2010). Emissions from main propulsion engine on container ship at sea, *Journal of Geophysical Research* D: Atmospheres, 115, 23.

[10] Jayaram, V., Nigam, A., Welch, W.A., Miller, J.W. and Cocker III, D.R. (2011). Effectiveness of emission control technologies for auxiliary engines on ocean-going vessels, *Journal of the Air and Waste Management Association*, 61, 1, 14-21.

[11] [11] Dalsøren, S.B., Endresen, O., Isaksen, I.S.A., Gravir, G. and Sörgärd, E. (2007). Environmental impacts of the expected increase in sea transportation, with a particular focus on oil and gas scenarios for Norway and northwest Russia, *Journal of Geophysical Research* D: Atmospheres, 112, 2.

[12] Kim, H.S., Song, C.H., Park, R.S., Huey, G. and Ryu, J.Y. (2009). Investigation of ship-plume chemistry using a newly-developed photochemical/dynamic ship-plume model, *Atmospheric Chemistry and Physics*, 9, 19, 7531-7550.

[13] Chen, G., Huey, L.G., Trainer, M., Nicks, D., Corbett, J., Ryerson, T., Parrish, D., Neuman, J.A., Nowak, J., Tanner, D., Holloway, J., Brock, C., Crawford, J., Olson, J.R., Sullivan, A., Weber, R., Schauffler, S., Donnelly, S., Atlas, E., Roberts, J., Flocke, F., Hübler, G. and Fehsenfeld, F. (2005). An investigation of the chemistry of ship emission plumes during ITCT 2002, *Journal of Geophysical Research* D: Atmospheres, 110, 10, 1-15.

[14] Deniz, C. and Durmuşo Lu, Y. (2008). Estimating shipping emissions in the region of the Sea of Marmara, Turkey, *Science of the Total Environment*, 390, 1, 255-261.

[15] Dalsøren, S.B., Eide, M.S., Myhre, G., Endresen, Ø., Isaksen, I.S.A. and Fuglestvedt, J.S. (2010). Impacts of the large increase in international ship traffic 2000-2007 on tropospheric ozone and methane, *Environmental Science and Technology*, 44, 7, 2482-2489.

[16] Petzold, A., Weingartner, E., Hasselbach, J., Lauer, P., Kurok, C. and Fleischer, F. (2010). Physical properties, chemical composition, and cloud forming potential of particulate emissions from a marine diesel engine at various load conditions, *Environmental Science and Technology*, 44, 10, 3800-3805.

[17] Wang, C., Corbett, J.J. and Winebrake, J.J. (2007). Cost-effectiveness of reducing sulfur emissions from ships, *Environmental Science and Technology*, 41, 24, 8233-8239.

[18] Wang, C., Corbett, J.J. and Firestone, J. (2007). Modeling energy use and emissions from North American shipping: Application of the ship traffic, energy, and environment model, *Environmental Science and Technology*, 41, 9, 3226-3232.

[19] Endresen, O., Bakke, J., Sørgård, E., Berglen, T.F. and Holmvang, P. (2005). Improved modelling of ship SO_2 emissions - A fuel-based approach. *Atmospheric Environment*, 39, 20, 3621-3628.

[20] Matthias, V., Bewersdorff, I., Aulinger, A. and Quante, M. (2010). The contribution of ship emissions to air pollution in the North Sea regions. *Environmental Pollution*, 158, 6, 2241-2250.

[21] Lack, D.A., Cappa, C.D., Langridge, J., Bahreini, R., Buffaloe, G., Brock, C., Cerully, K., Coffman, D., Hayden, K., Holloway, J., Lerner, B., Massoli, P., Li, S., McLaren, R.,

Middlebrook, A.M., Moore, R., Nenes, A., Nuaaman, I., Onasch, T.B., Peischl, J., Perring, A., Quinn, P.K., Ryerson, T., Schwartz, J.P., Spackman, R., Wofsy, S.C., Worsnop, D., Xiang, B. and Williams, E. (2011). Impact of fuel quality regulation and speed reductions on shipping emissions: Implications for climate and air quality. *Environmental Science and Technology*, 45, 20, 9052-9060.

[22] Wang, C. and Corbett, J.J. (2007). The costs and benefits of reducing SO_2 emissions from ships in the US West Coastal waters, Transportation Research Part D: Transport and Environment, 12, 8, 577-588.

[23] Petzold, A., Hasselbach, J., Lauer, P., Baumann, R., Franke, K., Gurk, C., Schlager, H. and Weingartner, E. (2008). Experimental studies on particle emissions from cruising ship, their characteristic properties, transformation and atmospheric lifetime in the marine boundary layer, *Atmospheric Chemistry and Physics*, 8, 9, 2387-2403.

[24] Lack, D.A., Corbett, J.J., Onasch, T., Lerner, B., Massoli, P., Quinn, P.K., Bates, T.S., Covert, D.S., Coffman, D., Sierau, B., Herndon, S., Allan, J., Baynard, T., Lovejoy, E., Ravishankara, A.R. and Williams, E. (2009). Particulate emissions from commercial shipping: Chemical, physical, and optical properties. *Journal of Geophysical Research* D: Atmospheres, 114, 4.

[25] Moldanová, J., Fridell, E., Popovicheva, O., Demirdjian, B., Tishkova, V., Faccinetto, A. and Focsa, C. (2009). Characterisation of particulate matter and gaseous emissions from a large ship diesel engine, *Atmospheric Environme*nt, 43, 16, 2632-2641.

[26] Winther, M. (2008). New national emission inventory for navigation in Denmark, *Atmospheric Environment*, 42, 19, 4632-4655.

[27] Agrawal, H., Welch, W.A., Miller, J.W. and Cocker, D.R. (2008). Emission measurements from a crude oil tanker at sea, *Environmental Science and Technology*, 42, 19, 7098-7103, 2008.

[28] Moldanová, J., Fridell, E., Popovicheva, O., Demirdjian, B., Tishkova, V., Faccinetto, A. and Focsa, C. (2010). *Characterisation of particulate matter and gaseous emissions from a large ship diesel engine*. 43, 16, 2632–2641.

[29] Lack, D., Lerner, B., Granier, C., Baynard, T., Lovejoy, E., Massoli, P., Ravishankara, A.R. and Williams, E. (2008). Light absorbing carbon emissions from commercial shipping, *Geophysical Research Letters*, 35, 13.

[30] Winnes, H. and Fridell, E. (2009). Particle emissions from ships: Dependence on fuel type, *Journal of the Air and Waste Management Association*, 59, 12, 1391-1398.

[31] Popovicheva, O., Kireeva, E., Shonija, N., Zubareva, N., Persiantseva, N., Tishkova, V., Demirdjian, B., Moldanová, J. and Mogilnikov, V. (2009). Ship particulate pollutants: Characterization in terms of environmental implication, *Journal of Environmental Monitoring*, 11, 11, 2077-2086.

[32] Ault, A.P., Gaston, C.J., Wang, Y., Dominguez, G., Thiemens, M.H. and Prather, K.A. (2010). Characterization of the single particle mixing state of individual ship plume events measured at the port of Los Angeles. *Environmental Science and Technology*, 44, 6, 1954-1961.

[33] Fitzgerald, W.B., Howitt, O.J.A. and Smith, I.J. (2011). Greenhouse gas emissions from the international maritime transport of New Zealand's imports and exports, *Energy Policy*, 39, 3, 1521-1531.

[34] Jonsson, A.M., Westerlund, J. and Hallquist, M. (2011). Size-resolved particle emission factors for individual ships, *Geophysical Research* Letters, 38, 13.

[35] Eide, M.S., Longva, T., Hoffmann, P., Endresen, Ø. and Dalsøren, S.B. (2011) Future cost scenarios for reduction of ship CO_2 emissions. *Maritime Policy and Management*, 38, 1, 11-37.

[36] Wang, H. (2010). Economic costs of CO_2 emissions reduction for non-Annex I countries in international shipping, *Energy for Sustainable Development*, 14, 4, 280-286.

[37] Heitmann, N. and Khalilian, S. (2011). Accounting for carbon dioxide emissions from international shipping: Burden sharing under different UNFCCC allocation options and regime scenarios. *Marine Policy*, 35, 5, 682-691.

[38] Song, C.H., Kim, H.S., Von Glasow, R., Brimblecombe, P., Kim, J., Park, R.J., Woo, J.H. and Kim, Y.H. (2010). Source identification and budget analysis on elevated levels of formaldehyde within the ship plumes: A ship-plume photochemical/dynamic model analysis. *Atmospheric Chemistry and Physics*, 10, 23, 11969-11985.

[39] Viana, M., Amato, F., Alastuey, A., Querol, X., Moreno, T., Dos Santos, S.G., Herce, M.D. and Fernández-Patier, R. (2009). Chemical tracers of particulate emissions from commercial shipping, *Environmental Science and Technology*, 43, 19, 7472-7477.

[40] Wang, H., Liu, D. and Dai, G. (2009). Review of maritime transportation air emission pollution and policy analysis, *Journal of Ocean University of China*, 8, 3, 283-290.

[41] Healy, R.M., O'Connor, I.P., Hellebust, S., Allanic, A., Sodeau, J.R. and Wenger, J.C. (2009). Characterisation of single particles from in-port ship emissions, *Atmospheric Environment*, 43, 40, 6408-6414.

[42] Miola, A. and Ciuffo, B. (2011). Estimating air emissions from ships: Meta-analysis of modelling approaches and available data sources. *Atmospheric Environment*, 45, 13, 2242-2251.

[43] Bovensmann, H., Eyring, V., Franke, K., Lauer, A., Schreier, M., Richter, A. and Burrows, J.P. (2006). Emissions of international shipping as seen by satellites, *European Space Agency*, (Special Publication) ESA SP.

[44] Endresen, Ø., Sørgård, E., Behrens, H.L., Brett, P.O. and Isaksen, I.S.A. (2007). A historical reconstruction of ships' fuel consumption and emissions, *Journal of Geophysical Research D: Atmospheres*, 112, 12.

[45] Franke, K., Eyring, V., Sander, R., Hendricks, J., Lauer, A. and Sausen, R. (2008). Toward effective emissions of ships in global models, *Meteorologische Zeitschrift*, 17, 2, 117-129.

[46] Corbett, J.J., Winebrake, J.J. and Green, E.H. (2010). An assessment of technologies for reducing regional short-lived climate forcers emitted by ships with implications for Arctic shipping. *Carbon Management*, 1, 2, 207-225.

[47] Paxian, A., Eyring, V., Beer, W., Sausen, R. and Wright, C. (2010). Present-day and future global bottom-up ship emission inventories including polar routes, *Environmental Science and Technology*, 44, 4, 1333-1339.

[48] Jalkanen, J., Brink, A., Kalli, J., Pettersson, H., Kukkonen, J. and Stipa, T. (2009). A modelling system for the exhaust emissions of marine traffic and its application in the Baltic Sea area. *Atmospheric Chemistry and Physics*, 9, 23, 9209-9223.

[49] Wang, C., Corbett, J.J. and Firestone, J. (2008). Improving spatial representation of global ship emissions inventories. *Environmental Science and Technology*, 42, 1, 193-199.

[50] Fridell, E., Steen, E. and Peterson, K. (2008). Primary particles in ship emissions. *Atmospheric Environment*, 42, 5, 1160-1168.

[51] Lack, D.A. and Corbett, J.J. (2012). Black carbon from ships: A review of the effects of ship speed, fuel quality and exhaust gas scrubbing. *Atmospheric Chemistry and Physics*, 12, 9, 3985-4000.

[52] Dedes, E.K., Hudson, D.A. and Turnock, S.R. (2012). Assessing the potential of hybrid energy technology to reduce exhaust emissions from global shipping. *Energy Policy*, 40, 1, 204-218.

[53] Duran, V., Uriondo, Z. and Moreno-Gutiérrez, J. (2012). The impact of marine engine operation and maintenance on emissions, *Transportation Research Part D: Transport and Environment*, 17, 1, 54-60.

SUBSECTION 4: PORT SHIP EMISSIONS

In: Ships and Shipbuilding
Editor: José A. Orosa

ISBN: 978-1-62618-787-0
© 2013 Nova Science Publishers, Inc.

Chapter 9

AIR QUALITY IMPACT ASSESSMENT OF IN-PORT SHIP EMISSIONS: METHODOLOGICAL ISSUES AND CASE-STUDY EXAMPLES

Giovanni Lonati[*]

Civil and Environmental Engineering Department
Politecnico di Milano, Como Province of Como, Italy

ABSTRACT

This chapter deals with the main steps of the impact assessment of port activities on air quality at the local scale, specifically focusing on the impact of emissions of ships while approaching and docking in port and while mooring at the quayside for offloading/loading operations. Methodological aspects of the assessment are discussed focusing on the ships' atmospheric emission estimation, on the definition of suitable emission scenarios for atmospheric dispersion modelling, on the set-up of model simulations and on the output concentration data processing for assessing compliance with both long- and short-term air quality standards near to the port. Basic principles of air dispersion modelling are also recalled, referring to both screening models, which provide maximum ground-level concentrations, and more refined dispersion models, which provide long-term average concentrations and depositions. The implementation of the methodological aspects is finally presented for two case-studies concerning the impact assessment on local air quality of in-port ship emissions for new freight and touristic ports in project.

INTRODUCTION

Maritime transportation is widely recognized as a relevant source of the total air pollution worldwide, with implications at both local and regional scale, other than at the global scale.

---

[*] Email: giovanni.lonati@polimi.it.

Shipping emissions are easily transferred for long distances in the atmosphere from sea to land and even from one continent to another [1]. Moreover, part of shipping emissions occur at coastal areas thus directly dispersing onto mainland and causing environmental problems and affecting human health, ecosystem and the environment [2]. Furthermore, ports located in proximity to urbanized area, produce a combined environmental effect due to the superposition of ship and port activity emissions with those related to urban sources [3, 4]. Well-known health effects associated with particulate matter (PM), ultrafine particles, ozone and NO_X at a local level comprise respiratory diseases and premature death from heart and pulmonary diseases.

Even if international shipping contributes by a small percentage of the total mass of atmospheric air pollution compared to road vehicles and other industrial sectors, the exhaust gas emissions (NO_X, SO_X, CO_X, HC, VOC and PM) are potentially affecting the environment by contributing to smog and acid rain and, in the case of carbon oxides, to the greenhouse effect. Emissions of trace pollutants like PCBs, PCDD/Fs and PAH have been also reported by [5, 6].

A number of research studies on the evaluation of the impact of atmospheric shipping emissions on air quality at the local scale and on climate at the regional scale have been reported in literature over the last decade. Some studies addressed the estimation of the magnitude of shipping emissions to the atmosphere [7-14]; other works addressed the characterisation of emissions from ships and the estimation of their impact at the local scale on urban areas proximal to harbour sites by means of dispersion modelling [15-19]. Some other works specifically addressed the identification and estimation of the impact of harbour activities on ambient PM levels in nearby urban areas, since ship emissions, loading, unloading and transport operations of dusty loose materials may have a significant impact on PM levels with subsequent potential effects on human health [20-22].

The two principal sources of air pollutants related to shipping transport are seagoing vessels and harbour operations: while the emissions from seagoing vessels mainly derive from combustion in main and auxiliary propulsion engines, harbour emissions derive from port activities as manoeuvring, loading/unloading operations and hotelling. Due to the potential environmental impact of port activities, the Environmental impact assessment (EIA) is often required for trading ports, harbours and port installations that by virtue of their nature, size or location are expected to have significant effects on the environment (Directive 85/337/EEC as amended by Directives [21] and [23]). Moreover, regardless of regulatory requirements, local communities where small ports (freight and/or tourist) are under project frequently ask for an environmental impact assessment, with major concerns related to air quality issues.

This chapter deals with the main steps of the impact assessment of port activities on air quality at the local scale, specifically focusing on the impact of emissions of ships while approaching and docking in port and while mooring at the quayside for offloading/loading operations. Methodological aspects of the main steps of the assessment are discussed focusing on the ships' atmospheric emission estimation, on the definition of suitable emission scenarios for atmospheric dispersion modelling, on the set-up of model simulations and on the output concentration data processing for assessing compliance with long-term (i.e.: annual average) and short-term (i.e.: daily and hourly averaged concentrations) air quality standards near to the port. In particular, issues related to the definition of base-case and worst-case emission scenarios and the assessment of the short-term impacts on air quality are specifically addressed. Basic principles of air dispersion modelling are also recalled, referring to both screening models, which provide maximum ground-level concentrations, and more refined dispersion models, which provide long-term average concentrations and depositions.

The implementation of the methodological aspects is finally presented for two case-studies concerning the impact assessment on local air quality of in-port ship emissions for new freight and tourist ports in project.

1. EMISSION ASSESSMENT

Vessel traffic is the first step for the subsequent emission assessment and atmospheric dispersion model simulations. This basic piece of information, concerning the vessel traffic data as number and fleet composition, engine features and operation, in-port residence time schedule, is usually available in the case of existing ports; conversely, it can only be estimated based on the main features of the facility in the case of a project for a new port. For pleasure yacht ports project data is typically the number of berths, for ferry ports the number of ferry lines and scheduled connection services, for freight ports the annual throughput of raw materials/products or the handling capacity of the containers; in this latter case the vessel traffic is determined based on further assumptions on the carrying capacity of the cargo ships.

The methodologies proposed in literature to estimate shipping emissions have been primarily developed in order to compile atmospheric pollutant emissions inventories. Basically there are two methods that can be used to produce emission estimates for navigation: the fleet-activity based method, which relies on detailed information on ship movements and ship classes, as well as the corresponding fuel consumption figures and emission factors, and the fuel based method, which relies on fuel sales data in combination with fuel-related emission factors. However, given the limited spatial scale of interest, the fleet-activity based method is more suitable for emission assessment in the framework of a local-scale EIA study, as suggested by [24], who recently proposed a fleet-activity based method for local, regional and national shipping emissions assessment combined with a fuel-based method for international shipping.

The most important European fleet-activity based methodologies have been developed in the framework of the MEET project [25], of the TREND system [26] and by ENTEC [27]; this type of methodology has been applied in other works [12, 28] and a new reference system for emission factors, energy consumption, and total emissions for maritime sources has been recently developed within the European EX-TREMIS project [29].

The fleet-activity based methodologies for estimating shipping emissions commonly consider four different navigation phases or "modes": a) approaching and docking in port; b) hotelling (or berthing) in port; c) departing from the port; and d) cruising. The manoeuvring phase, which includes modes a) and c), is assumed to start when the ship's deceleration begins and to end at the moment of docking, recommencing with the departure from berth and then ending when cruising speed has been reached. The hotelling in port phase spans the time the ship is at the dockside with its main engines working at reduced load to generate power in order to supply the ship's main onboard services (lighting, heating, refrigeration, ventilation systems). However, ships powered by internal combustion engines normally use auxiliary engines or diesel powered generators to supply auxiliary power; alternative option that prevents hotelling emissions is shore electricity supply. For liquid bulk ships power requirements also include the power of the cargo pumps for tanker off-loading and that of the ballast pumps for tanker loading; as these latter power requirements can be relatively high, the related emissions are estimated separately.

Emission factors used in fleet-activity based methodologies are typically expressed in terms of mass of pollutant per unit of engine power or in terms of mass of pollutant per unit mass of fuel, referred to the different navigation modes, hence regarded as modal emission factors, for given fuel and engine. However, for in-port emission assessment, especially at the project stage for a new port, it is easier to use emission factors in terms of mass of pollutant per unit mass of fuel, since actual load factor and operational time of ships' engines is a too refined piece of information in comparison with fuel consumption.

As a general approach the overall emissions E of an atmospheric pollutant during navigation are estimated as:

$$E = \sum_m{}' AF_m \cdot EF_m \cdot t_m \tag{1}$$

where AF_m is a modal activity factor (i.e.: actual fuel consumption rate or actual engine power used in the navigation mode), EF_m is the modal emission factor expressed in units coherent with AF (i.e.: mass of pollutant per unit mass of fuel or per unit engine power and time), and t_m is the time spent in each navigation mode over the reference period.

According to the fleet-activity based MEET methodology the general equation (1) to assess the overall emission E_i of pollutant i takes the following formulation:

$$E_i = \sum_{jklm} n_{jkl} \cdot E_{ijklm} = \sum_{jklm} n_{jkl} \cdot \left[S_{jklm}(GT) \cdot t_{jklm} \cdot F_{ijlm} \right] \tag{2}$$

where:

- j is the fuel used (i.e.: bunker fuel oil, marine diesel oil, marine gas oil, gasoline fuel);
- k is the ship class for use in fuel consumption classification (i.e.: Solid Bulk, Liquid Bulk, General Cargo, Container, Passenger/Ro-Ro/Cargo, Passenger, High speed ferries, Inland Cargo, Sail ship, Tugs, Other);
- l is the engine type class for use in emission factors characterization (i.e.: steam turbines, high, medium, or slow speed motor engines);
- m is the navigation mode (i.e.: cruising, manoeuvring, hotelling, tanker offloading, auxiliary generators).
- n_{jkl} is the number of ships from class k using fuel j in type l engines in transit;
- E_{ijklm} is the emissions of pollutant i from one class k ship using fuel j in type l engines during navigation mode m;
- S_{jklm} is the actual daily consumption of fuel j in ship class k in mode m as a function of gross tonnage (GT);
- t_{jklm} is the time in navigation mode m of ships of class k with engines type l using fuel j;

- F_{ijlm} are the modal emission factors of pollutant i from fuel j in engines type l in mode m.

In particular, when assessing in-port ship emissions eq. (2) has to be applied considering for the m index the manoeuvring, hotelling, and auxiliary engine phase, eventually.

The actual daily fuel consumption S_{jklm} can be calculated based on the daily fuel consumption $C_{jk}(GT)$ of fuel j in ship class k at full power and on the fraction p_m of the full-power fuel consumption corresponding to the navigation mode m; in turn, the full-power fuel consumption $Cjk(GT)$ can be estimated by means of empirical relationships as a function of gross tonnage GT.

As far as in-port emissions are concerned the time in navigation mode m normally accounts for both the manoeuvring and the real hotelling at berth: however, time spent on manoeuvring is typically only 1-2 hours while time spent at berth can be one or more days, depending on several factors, mainly related to the duration of loading and unloading operations. Average hotelling time and fuel consumption rates at berth for different ship types in the harbour of Rotterdam are reported [30].

However, during the manoeuvring and hotelling modes ship can use both main engines and auxiliary engines with different loads: in the work of Mölders [13] it is assumed that during manoeuvring ships use 20% of the main and 50% of the auxiliary engine load whilst during berthing ships, except tankers, only use the auxiliary engines at 60% load; for tankers a 20% and 60% load was assigned to the main and auxiliary engines, respectively.

For the assessment of emission from auxiliary engines used to generate electricity onboard of berthing ships a simpler approach is proposed by the *Emission estimation technique manual for maritime operations* issued by the Department of Sustainability, Environment, Water, Population and Communities of the Australian Government [31]. This approach is directly based on the auxiliary engine power, on the hotelling time and on emission factors expressed as pollutant mass flow rate per unit power of the engine; further simplifications are the assumption of the same auxiliary engine power (600 kW) and the use of an average hotelling time for all the ships.

For the emission from the boilers that most vessels have to heat residual oil to make it fluid enough to use in diesel engines and to produce hot water [32] the assessment is more frequently based on fuel consumption rate than on the thermal power of the boiler, since emission factors are usually given in terms of fuel usage, rather than power.

If the fuel consumption rate is unknown, it is recommended that a fuel consumption rate of 0.0125 metric tons of fuel per hour is used to estimate emissions from auxiliary boilers [32]. Emission factors to estimate emissions from auxiliary engines and boilers are provided in the manual for each fuel type; calculation examples for emissions from auxiliary engines and boilers are also given in the manual.

Emission factors for the different navigation modes, ship and engine class, fuel are available from ENTEC [27] for NO_X, SO_2, VOCs and PM, from [6], [7] for CO, from [6] for NH_3, from [33] for NO_X, SO_X, primary PM, CO and VOC, from [30] for hydrocarbons, NO_X, SO_2, CO and PM10.

In general, port area emissions do not derive from manoeuvring/berthing vessels only, but also from a number of other sources such as harbour craft (tugs, pilot boats, dredgers), tank fuel storage, bulk dry material handling for transfer (i.e. conveyors to conveyors, conveyor to stockpile, front hand loaders), bulk volatile material loading, landside vehicle operation, maintenance operations, training fires.

Methodologies and examples for the assessment of such emissions are reported by ENTEC [27], by CORINAIR [34], and in the works [35]. Procedures and recommended approaches for estimating emissions from facilities engaged in maritime operations, namely those primarily engaged in the operation of ports (i.e.: loading and unloading of freight), ballasting, transit, and maintenance and general upkeep of vessels, are also thoroughly described in the abovementioned manual from [31].

2. ATMOSPHERIC DISPERSION MODELS

Atmospheric dispersion modelling at coastal sites is rather challenging, essentially due to the heterogeneity in meteorological and dispersion conditions generated by the abrupt change of the surface features at the coastline. Spatial and temporal variability of the internal boundary layer [36], sea-land breeze circulations and shoreline fumigation are some of the factors to be taken into account, thus suggesting the use of advanced dispersion models. Literature studies report the use of both advanced Gaussian dispersion models [16, 18], puff models [37] and Lagrangian particle dispersion models [17].

One of the key elements of an effective dispersion modelling study is to choose an appropriate tool to match the scale and significance of potential impact and the complexity of dispersion, essentially regulated by terrain and meteorology effects. Several dispersion models have been developed: some simplified models (screening models) are used before applying a refined air quality model to determine whether refined modelling is needed. Some models are considered as regulatory models by the national Environmental Protection Agencies; some others are considered as advanced models to be used for complex atmospheric and orographic situations. Compilations listing the most common dispersion models can be easily found on the web (for instance: U.S. EPA SCRAM website: www.epa.gov/scram001 or Wikipedia List of atmospheric dispersion models: http://en.wikipedia.org/wiki/List_of_ atmospheric_dispersion_models).

2.1. Box Model

Though conceptually simple, the box model is suitable for application to the impact assessment due to its conceptual approach for the emission source description and to the small amount of input meteorological data [38-40]. Under conditions where pollutants are emitted over a large area and the assessment of area-wide average concentrations is required, the well-mixed box model approach provides a simple solution for the atmospheric dispersion. The model is based upon the mass balance of the pollutant over the box, stating that the accumulation of the pollutant within the box is balanced by input from upwind sources and output from the downwind edge of the box.

Box modelling is typically applied when it can be assumed that the concentration over the study area is approximately uniform as a consequence of the high number of single-point sources or because the emission source can be considered as an area source. In general, the extent of the box corresponds to the area of the source itself, whereas the vertical extent is assumed to be the atmospheric mixing layer. Basic assumption in box modelling is that the pollutant is uniformly mixed in the lower atmospheric layer between the ground and the mixing layer height.

2.2. Gaussian Dispersion Model

The Gaussian plume dispersion model equation is at the core of most of the screening and dispersion models, such as SCREEN ([41] and ISC3 [42], AERSCREEN [43] and AERMOD [44], CTDMPLUS [45], and AUSPLUME [46], and of advanced models such as CALPUFF [47] and TAPM [48].

The basic Gaussian dispersion model equation is derived by the mass balance for a specific pollutant in a control volume according to an Eulerian approach, which describes the changes of the pollutant concentration in the atmosphere as they occur at a fixed point in a 3D reference system. Transport mechanisms involved in the mass balance are advection by wind and turbulent diffusion, whereas molecular diffusion is neglected as considered negligible with respect to the other two mechanisms. An analytical solution of the mass balance equation can be obtained for the steady-state transport and diffusion problem of a passive pollutant over an idealized flat terrain by means of some simplifying assumptions.

However, in situations of complex terrain or near coastal boundaries, significant changes in meteorological conditions can occur over short distances due to sea breezes or slope and valley flows or other meteorological phenomena. In these cases, advanced models can simulate pollutant transport and dispersion in a much more realistic way than a Gaussian-plume model, which assumes spatial uniformity in the meteorology. Clearly, this means that advanced models require more detailed meteorological input data to accurately emulate the complex dispersion effects.

3. CASE STUDIES

Two case-studies examples of impact assessment on local air quality due to in-port ship emissions are presented below: the former case-study refers to the project for a new freight port the latter to the project of enlargement of an existing touristic port.

3.1. Case-Study #1: Freight Port

This case-study example refers to the revamping project for an existing small port in Southern Italy. The new port is intended to serve a new coal-fired power plant located on the sea front for the supply of coal and of all the other materials needed for plant operation, as well as the removal of residuals for their final disposal, exclusively by sea.

The impact on local air quality of in-port ship emission has been assessed by first estimating manoeuvring and hotelling phase emissions and then by atmospheric dispersion modelling, which provides the spatial distribution of pollutant concentrations in the study area, enabling for the impact assessment evaluation both in terms of long-term (i.e.: annual average) and short-term (i.e.: daily and hourly averaged concentrations) expected concentration levels. Details of the assessment steps are reported [19].

3.1.1. Emission Assessment

For the emission assessment the updated MEET methodology [33] was followed, since most suitable to estimate shipping emissions in harbour given the degree of knowledge of the harbour's activities, limited to a likely estimated number of vessels and related gross tonnage. The estimated ship traffic was about a hundred ships per year and included two main ship categories: Solid bulk carriers (SBC), large cargo ships (Panamax and Capesize class ships) used to transport coal, and General Cargos (GC), smaller merchant vessels used for transport of other raw materials and of the residuals of the coal combustion process. It was assumed that SBCs use Bunker Fuel Oil (BFO) with a 2.7% sulphur content as fuel, while GCs use Marine Diesel Oil (MDO), with a 0.5% sulphur content; further assumption was that both ship categories use auxiliary diesel engines fuelled with MDO while in hotelling phase.

Eq. (2) was applied to assess the annual emissions of NO_X, SO_X, PM, CO and VOC: the full-power daily fuel consumption $C_{jk}(GT)$ has been calculated by means of the empirical relations reported in Table 1; subsequently, the actual daily consumption in mode m has been calculated based on the full-power consumption values and on the p_m factors (0.4 and 0.2 for the for the manoeuvring and hotelling phase, respectively). Concerning the duration of the navigation modes, manoeuvring time has been hypothesized on the basis of literature data and port authorities' information as proportional to the size of the ship, since related to the difficulty in moving while approaching and leaving the port. The hotelling time has been estimated as the time required for ship offloading/loading, evaluated based on the ships' tonnage and on the offloading/loading systems' capacity, also considering eventual overnight stay at berth. It was assumed that SBCs are offloaded by continuously operating systems with shift works set on 24 hours per day, while for GCs offloading/loading operations were set on 3-shift works for 18 hours of work per day. Table 2 reports the emission factors depending on engine type, fuel used and operating mode whereas all ship data used for the emissions assessment are summarised in Table 3.

Table 1. Case-study #1: formulations for full-power fuel consumption estimate C_{jk} (kg day^{-1})

Ship class	Full-power daily fuel consumption
Solid bulk	$C_{jk} = 12.0724 + 0.0012 \cdot GT - 1.1501 \cdot 10^{-8} \cdot GT^2 + 4.6484 \cdot 10^{-14} \cdot GT^3$
General cargo	$C_{jk} = -2.2602 + 0.0049 \cdot GT - 1.6401 \cdot 10^{-7} \cdot GT^2 + 1.7394 \cdot 10^{-12} \cdot GT^3$

Table 2. Case-study #1: Emission factors (kg Mg_{fuel}^{-1})
(°) SO_X emission factor is given by: 20·Sulphur content in fuel (%)

Phase	Engine type – Fuel	NO_x	CO	VOC	PM	SO_x(°)
Manoeuvring	Medium speed diesel engine – MDO	51	28	3.6	1.2	10
	Slow speed diesel engine – BFO	78	28	3.6	7.6	54
Hotelling	Diesel auxiliary engines – MDO	63	4	1.9	1.1	10

Air Quality Impact Assessment of in-Port Ship Emissions

Table 3. Case-study #1: vessel traffic data and ship features

	Mode m	SBC #1	SBC #2	GC #1	GC #2	GC #3	GC #4	GC #5
Class k		Solid bulk SBC		General cargo GC				
Fuel j	Manoeuvring	Bunker Fuel Oil (BFO)		Marine diesel oil (MDO)				
	Hotelling	Marine diesel oil (MDO)		Marine diesel oil (MDO)				
Engine l	Manoeuvring	Slow speed diesel engine		Medium speed diesel engine				
	Hotelling	Diesel auxiliary engine		Diesel auxiliary engine				
p_m factor	Manoeuvring	0.40		0.40				
	Hotelling	0.20		0.20				
GT (Mg)		80000	130000	20000	5000	10000	10000	6500
t_{jklm} (days)	Manoeuvring	0.08	0.08	0.06	0.04	0.04	0.04	0.04
	Hotelling	2.22	3.61	2.17	0.42	1.08	1.08	0.54
Ships year^{-1}		14	16	12	23	11	7	10

3.1.2. Atmospheric Dispersion Modelling

The assessment of the ground level spatial distribution of atmospheric pollutants emitted by ships was performed by means of the CALMET/CALPUFF modelling system [47] for reference year 2004. CALPUFF is a multi-layer, non-steady-state Lagrangian Gaussian puff model containing modules for complex terrain effects, overwater transport, coastal interaction effects, building downwash, wet and dry removal and simple chemical transformation. Input meteorological data for the CALPUFF model are the 2D and 3D fields of the main local meteorological parameters (such as wind speed and direction, turbulence and atmospheric stability parameters, temperature, mixing layer height, precipitation rate) generated by the CALMET meteorological pre-processor, able to simulate also local effects like slope flows, kinematic terrain effects and sea breeze circulations. Input data for CALMET were the ground-level conventional meteorological data measured at the port site and both ground-level and vertical stratified data from large scale meteorological model simulation.

CALPUFF modelling domain was a study area centred on the port area and extending for 3.5 km in longitude and for 2.5 km in latitude. Concentrations have been calculated at the 936 nodes of a 36x26 Cartesian receptor grid with 100 m node spacing and at 6 discrete receptors located at the main residential nuclei of the area; terrain elevation in the study area was from sea-level up to about 250 m.

For model simulation purpose, SBCs and GCs estimated emissions have been separately considered: two area sources have been defined according to the different locations of the offloading/loading docks for SBC and GC in the port area. Given their size, in fact, GCs can manoeuvre inside the port and reach the inner dock for loading/offloading operations, whereas SBCs can only moor at the sea-front dock during coal offloading operations. The emission area for GCs was thus corresponding to the port surface (extending for about 211000 m^2), while the emission area for SBCs (extending for about 10000 m^2) was located in correspondence with their docking area on the seafront.

3.1.3. Model Simulations Set-Up

For small ports as in this case-study, the port operation is not continuous as a consequence of both the small number of vessels and their short hotelling time, leading to important implications for the model simulation set-up.

In this case-study the air quality impact assessment was performed with respect to different emission scenarios intended to produce representative long-term (i.e.: annual average) and short-term (i.e.: daily and hourly averaged concentrations) expected concentration levels in the study area. In fact, though the CALPUFF model can deal with time varying emissions, any assumption about the time pattern of vessel traffic would be arbitrary and would lead to results absolutely determined by the assumption itself. In particular, in order to assess annual average concentrations an emission scenario based on total hotelling emissions was considered, assuming that emission are uniformly distributed during the entire year. For daily and hourly average concentrations worst-case scenarios aiming to assess maximum short-term concentrations have been considered, based on actual hotelling emissions of the main emitters ships for the ship class SBC and GC (Capesize and GC#1). Moreover, a further scenario was set considering concurrently hotelling emissions of SBC Capesize and of GC #1, since on a busy day only 1 SBC and 1 GC can be simultaneously present for port operation.

3.1.4. Results

The most relevant emissions were estimated for NO_X, with a total emission of about 111 Mg year^{-1}, followed by SO_X with roughly 20 Mg year^{-1}; less relevant from the quantitative standpoint were CO emissions (10 Mg year^{-1}) and especially those of VOC and PM (about 3.5 and 2.5 Mg year^{-1}, respectively). For all the pollutants estimated emissions during the hotelling phase were largely prevailing on those of the manoeuvring phase, given its longest duration especially in the case of SBCs and GC#1: more than 90% of NO_X were estimated from the hotelling phase. Therefore, since manoeuvring emissions mainly occurred at sea during the approach and departure from the port, atmospheric dispersion modelling and air quality impact assessment only considered at-berth emission at the sea-front dock for SBCs and in the port for GCs. The estimated additional contributions to the annual average concentrations from vessel traffic emissions were not such to result in air quality limits non-attainment, given the rather low actual concentration levels in the area. Maximum annual average concentration was estimated at a grid node located at sea just in front of the port area at a distance of about 250 m from the shoreline (Figure 1). NO_2 annual average concentration was about 16 $\mu g\ m^{-3}$, those of SO_2 and CO were in the orders of a few $\mu g\ m^{-3}$ (3.5 $\mu g\ m^{-3}$ for SO_2 and 1.4 $\mu g\ m^{-3}$ for CO, respectively) whereas PM and VOC concentrations were below 1 $\mu g\ m^{-3}$, respectively around 0.43 and 0.67 $\mu g\ m^{-3}$. At the discrete receptors points annual average concentrations were on the average one order of magnitude lower than the area maximum value, with the highest concentration estimated for Receptor #1, located at the small village closest to the port area. Worst-case scenario analyses for 1-h average concentration pointed out that non-compliance with NO_2 air quality limits could occur at some receptors whilst non-attainment for SO_2 limit was not expected given the lower emissions and the highest reference threshold (350 $\mu g\ m^{-3}$) compared to NO_2 (200 $\mu g\ m^{-3}$).

Figure 1. Case-study #1: NO_2 annual average concentration ($\mu g\ m^{-3}$) contour plot.

The estimated number of exceedances of the reference threshold for 1-h NO_2 concentration at the discrete receptors is separately reported in Table 4 for SBC Capesize and GC#1 and for the busy day scenario.

Receptor #1 was the most exposed discrete receptor: the air quality limit (maximum 18 1-h exceedances of the reference threshold per year) was by far not attained in all of the scenarios; non-attainment was also expected at Receptor #2, as a consequence of SBC capsize emissions and on busy day scenario.

Despite the very conservative approach, since basic assumption was that ships were at-berth during the worst periods from the meteorological standpoint, these results suggested to supply shore electricity to the long-berthing ships as a viable solution in order to prevent most of the hotelling emissions, so that very limited additional impacts on air quality could derive from in-port ship emissions.

Table 4. Case-study #1: NO_2 maximum annual number of exceedances of 200 $\mu g\ m^{-3}$ for 1-h average concentration in worst-case scenario

Ship class	Discrete receptor						Air quality limit
	#1	#2	#3	#4	#5	#6	
SBC Capesize	42	21	11	4	8	8	
GC #1	33	7	1	0	0	0	18
SBC Capesize and GC #1	70	24	13	6	8	10	

3.2. Case-Study #2: Tourist Port

This case-study example refers to the enlargement project for an existing tourist port in Central Italy. The new port is intended for sail boats and motor boats measuring between 8 and 40 metres, as well as fishing boats measuring up to 18 metres, tourist motor boats and ferry service boats, for a total of approximately 800 boat slips with fingers and piles. The assessment of boats' emissions was intended to evaluate the long-term impact on NO_X concentration levels in a naturalistic area close to the port site. Given this aim, a screening model for atmospheric dispersion calculation has been used.

3.2.1. Emission Assessment

Emissions have been estimated for two scenarios representing the current and the prospected port configuration. Conversely to the case-study #1, in this case-study the emissions are almost exclusively associated to boats' manoeuvre in the port area, given the tourist feature of the port. Estimation was performed through eq. (1) based on the engine power used in the manoeuvring mode, on the corresponding emission factors, and on the manoeuvring time for in-port movements in the current and prospected port configuration.

Boat traffic was estimated as a fraction of the total number of slips available in port (about 25%), derived from observation of the current port activity. As a result, a larger number (almost double) of boats' movements was expected in the enlargement project scenario but, however, with a shorter overall manoeuvring time (25% less): therefore, as a combined result, the annual NO_X emission was expected to increase by only about 50%.

3.2.2. Atmospheric Dispersion Modelling

The assessment of the ground level spatial distribution of atmospheric pollutants emitted by boats in the port area was performed by means of the ISC3 model [42]. ISC3 model is a steady-state Gaussian plume model which can be used to assess pollutant concentrations from a wide variety of sources; formerly preferred air dispersion model in the US EPA's "Guideline on Air Quality Models", ISC3 is now regarded as an alternative model for atmospheric dispersion modelling analyses.

Input meteorological data for the ISC3 model are the time series of the main local meteorological parameters (wind speed and direction, atmospheric stability parameters, temperature, mixing layer height, precipitation rate) measured at or close to the port site.

Actually, ISC3 model was applied in its long-term version for NO_X annual average concentration prediction. This version of the model requires input meteorological data as joint-frequency distributions of wind speed, wind direction, and atmospheric stability classification, obtained from long-period observations.

ISC3 modelling domain was a study area centred on the port area and extending for 2.5 km in longitude and for 3.0 km in latitude. Concentrations have been calculated at the 806 nodes of a 36x21 Cartesian receptor grid with 100 m node spacing; terrain elevation was not accounted for since the study area is a flat area at sea-level.

For model simulation purpose, in both scenarios the port area has been regarded as an area source with a uniform NO_X emission rate over the entire port surface, extending for about 20000 m^2 in the current configuration and for about 115000 m^2 according to the port enlargement project.

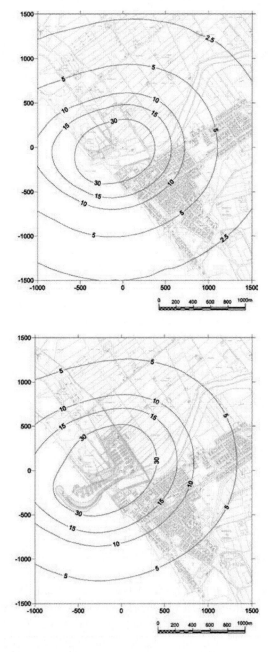

Figure 2. Case-study #2: NO_X annual average concentration ($\mu g\ m^{-3}$) contour plots. Current scenario (top panel), future scenario (bottom panel).

3.2.3. Model Simulations Set-Up

In this case-study the air quality impact assessment was intended to provide screening results about the NO_X annual average concentration, with special regard for the sea front area North to the port site due to its naturalistic and environmental value, and to allow the comparison between the expected impacts for the current and future scenario.

As in case-study #1, port operation can be hardly defined; therefore, broad assumptions were required for the model simulation set-up. In particular, with a very conservative approach, it was assumed that the number of boats' movements and atmospheric emissions as a straightforward consequence were constant throughout the entire year. This approach, applied to both scenarios, was intended to produce an upper bound estimate of the annual average concentration values, in agreement with the screening purpose of the impact assessment.

3.2.4. Results

Following to the port enlargement NO_X emissions were expected to grow by 50% in the future scenario, up to about 120 metric tons per year in the conservative assessment for the future scenario.

However, due to the significant port area enlargement the daily emission per unit area were estimated to decrease from about 10 g m^{-2} of the current scenario to about 2.5 g m^{-2} in the future scenario. The spatial distributions of the estimated contribution of port operation to the NO_X annual average concentration (μg m^{-3}) resulting from the model simulations are presented in Figure 2 for both the current scenario and future scenario.

The port area enlargement partially offset the effect of the increased emissions in the future scenario so that no significant change in the current NO_X levels were expected following to the port enlargement with estimated concentration levels were within the 5-30 μg m^{-3} range in both scenarios, in substantial agreement with the critical levels for the protection of vegetation set by current regulations.

Despite the lower emission rate per unit surface expected in the future scenario, the area impacted by port emissions is somewhat enlarged, with the concentration isophlets extending over a slightly larger area, still keeping the typical round shape characteristic of large ground-level area sources.

REFERENCES

[1]	Li, Q., Jacob, D. J., Bey, I., Palmer, P. I., Duncan, B. N., Field, B. D., Martin, R. V., Fiore, A. M., Yantosca, R. M., Parrish, D. D., Simmonds, P. G., Oltmans, S. J. (2002). Transatlantic transport of pollution and its effects on surface ozone in Europe and North America. *Journal of Geophysical Research*, 107(D13), 4166.

[2]	Lu, G., Brook, J.R., Rami Alfarra, M., Anlauf, K., Leaitch, W. R., Sharma, S., Wang, D., Worsnop, D.R., Phinney, L. (2006). Identification and characterization of inland ship plumes over Vancouver, BC. *Atmospheric Environment*, 40, 2767-2782.

[3]	Mueller, D., Uibel, S., Takemura, M., Klingelhoefer, D., Groneberg, D. A. (2011). Ships, ports and particulate air pollution – an analysis of recent studies. *Journal of Occupational Medicine and Toxicology*, 6, 31.

[4]	Schembari, C., Cavalli, F., Cuccia, E., Hjorth, J., Calzolai., G., Pérez, N., Pey, J., Prati, P., Raes, F. (2012). Impact of a European directive on ship emissions on air quality in Mediterranean harbours, *Atmospheric Environment,* 61, 661-669.

[5]	Cooper, D.A. (1993). Exhaust emissions from ships at berth. *Atmospheric Environment*, 37, 3817-3830.

[6]   Cooper, D. A., Gustafsson, T. (2004). Methodology for calculating emissions from ships: 1. Update of Emission factors. *Report series for SMED and SMEDandSLU,* Swedish Meteorological and Hydrological Institute, Norrköping, Sweden.

[7]   Corbett, J.J., Fischbeck, P. (1997). Emissions from ships. *Science,* 278 (5339), 823-824.

[8]   Corbett, J.J., Koehler, H.W. (2003). Updated emissions from ocean shipping. *Journal of Geophysical Research,* 108 (D20), 4650.

[9]   Endresen, Ø., Sørgård, E., Sundet, J.K., Dalsøren, S.B., Isaksen, I.S.A., Berglen, T.F., Gravir, G. (2003). Emission from international sea transportation and environmental impact. *Journal of Geophysical Research* 108, 4560.

[10] Deniz, C., Durmuşoğlu, Y., (2007). Estimating shipping emissions in the region of the Sea of Marmara, Turkey. *Science of the total environment,* 390, 255-261.

[11] Janhäll, S. (2007). Particle Emissions from Ships. The alliance for global sustainability, *Chalmers university,* Göteborg, Sweden, ISBN: 978-91-976534-3-5.

[12] De Meyer P., Maes, F., Volckaert, A., (2008). Emissions from international shipping in the Belgian part of the North Sea and the Belgian seaports. *Atmospheric Environment,* 42, 196-206.

[13] Mölders, N., Porter, S.E., Cahill, C.F., Grell, G.A. (2010). Influence of ship emissions on air quality and input of contaminants in southern Alaska National Parks and Wilderness Areas during the 2006 tourist season. *Atmospheric Environment,* 44, 1400-1413.

[14] Tzannatos, E. (2010). Ship emissions and their externalities for the port of Piraeus – Greece. *Atmospheric Environment,* 44, 400-407.

[15] Isakson, J., Persson, T.A., Selin Lindgren, E. (2001). Identification and assessment of ship emissions and their effects in the harbour of Göteborg, Sweden. *Atmospheric Environment* 35, 3659-3666.

[16] Saxe, H., Larsen, T. (2004). Air pollution from ships in three Danish ports. *Atmospheric Environment,* 38, 4057-4067.

[17] Gariazzo, C., Papaleo, V., Pelliccioni, A., Calori, G., Radice, P., Tinarelli, G. (2007). Application of a Lagrangian particle model to assess the impact of harbour, industrial and urban activities on air quality in the Taranto area, Italy. *Atmospheric Environment,* 41, 6432-6444.

[18] Lucialli, P., Ugolini, P., Pollini, E. (2007). Harbour of Ravenna: The contribution of harbour traffic to air quality. *Atmospheric Environment,* 41, 6421-6431

[19] Lonati, G., Cernuschi, S., Sidi, S. (2010). Air quality impact assessment of at-berth ship emissions: case-study for the project of a new freight port. *Science of the total environment,* 409, 192-200.

[20] Sharma, D.C. (2006). Ports in storm. *Environmental Health Perspectives,* 114, 4.

[21] European Union Council Directive 97/11/EC of 3 March 1997 amending Directive 85/337/EEC on the assessment of the effects of certain public and private projects on the environment. *Official Journal* L 73, 14.3.1997.

[22] Alastuey, A., Moreno, N., Querol, X., Viana, M., Artíñano, B., Luaces, J.A., Basora, J., Guerra, A. (2007). Contribution of harbour activities to levels of particulate matter in a harbour area: Hada Project-Tarragona Spain. *Atmospheric Environment,* 41, 6366-6378.

[23] European Union Directive 2003/35/EC of the European Parliament and of the Council of 26 May 2003, providing for public participation in respect of the drawing up of certain plans and programmes relating to the environment and amending with regard to

public participation and access to justice Council Directives 85/337/EEC and 96/61/EC. *Official Journal L* 156, 25.6.2003.

[24] Winther, M., (2008). New national inventory for navigation in Denmark. *Atmospheric Environment*, 42, 4632-4655.

[25] Trozzi, C., Vaccaro, R. (1998). Methodologies for estimating air pollutant emissions from ships. *Techne report MEET*, 1998.

[26] Georgakaki A., Coffey, R.A., Lock, G. Sorenson, S.C. (2004). Transport and Environmental Database System (TRENDS): Maritime air pollution emission modelling. *Atmospheric Environment*, 39, 2357-2365.

[27] ENTEC UK Ltd. (2002). Quantification of emissions from ships associated with ship movements between ports in the European Community. Final Report.

[28] Schrooten, L., De Vlieger, I., Int Panis, L., Styns, K., Torfs, R. (2008). Inventory and forecasting of maritime emissions in the Belgian sea territory, an activity-based emission model. *Atmospheric Environment*, 42, 667-676.

[29] Schrooten, L., De Vlieger, I., Int Panis, L., Chiffi, C., Pastori, E. (2009). Emissions of maritime transport: A European reference system. *Science of the total environment*, 408, 318-323.

[30] Hulskotte, J.H.J., Denier van der Gon, H.A.C. (2010). Fuel consumption and associated emissions from seagoing ships at berth derived from an on-board survey. *Atmospheric Environment*, 44, 1229-1236.

[31] Australian Government, Department of Sustainability, Environment, Water, Population and Communities. (2012). Emission Estimation Technique Manual for Maritime Operations Version 2.1, July 2012. ISBN 978 0 642 55425 3 available at http://www.npi.gov.au /publications/emission-estimation-technique/fmaritime.html.

[32] ICF Consulting. (2006). Current Methodologies and Best Practices in Preparing Port Emission Inventories. EPA, Fairfax, Virginia, USA. Final Report for U.S. EPA, January 2006.

[33] Trozzi, C., Vaccaro, R. (2006). Methodologies for estimating air pollutant emissions from ships: a 2006 update. Proceedings 2nd Conference Transport and Air Pollution, Reims, France, *Actes INRETS* 107, Vol 2.

[34] CORINAIR, (2006). Shipping activities. *Emission Inventory Guidebook*, B842 1-24.

[35] Peng, C., Lin, C., Jong, T. (2005). Emissions of particulate and gaseous pollutants within the Keelung harbour region of Taiwan. *Environmental Monitoring and Assessment*, 109, 37-56.

[36] Batchvarova, E., Xiaoming, C., Gryning, S. E., Steyn, D. (1999). Modelling internal boundary-layer development in a region with a complex coastline. *Boundary-layer meteorology*, 90, 1, 1-20.

[37] Indumati, S., Oza, R. B., Mayya, Y. S., Puranik, V.D., Kushwaha, H. S. (2009). Dispersion of pollutants over land–water–land interface: Study using CALPUFF model. *Atmospheric Environment*, 43, 473-478

[38] Lettau, H. (1970). Physical and meteorological basis for mathematical models of urban diffusion. Proceedings Symposium on MultipleSource Urban Diffusion Models. *Air Pollution Control Official Publication* No. AP 86, Environmental Protection Agency.

[39] Hanna, S. R. (1971). A simple method of calculating dispersion from urban area sources. *Journal of the Air Pollution Control Association*, 21, 774-777.

[40] Hanna, S.R., Briggs, G.A., Hosker, R.P. Jr. (1982). Handbook on Atmospheric Diffusion, U.S. Department. of Energy, DOE TIC-11223. http://dx.doi.org/10.2172/5591108.

[41] U.S. EPA, (1995a). SCREEN3 User's Guide. U.S. EPA Report EPA-454/B-95-004.

[42] U.S. EPA, (1995b). User's Guide for the Industrial Source Complex (ISC3) Dispersion Model. Volume II – Description of Model Algorithms. U.S. EPA Report EPA-454/B-95-003B.

[43] U.S. EPA, (2011). AERSCREEN User's Guide. U.S. EPA Report EPA-454/B-11-001, March 2011.

[44] Cimorelli, A.J. et al., (2004). AERMOD: Description of Model Formulation. U.S. EPA report EPA-454/R-03-004.

[45] Perry, S.G., et al. (1989). User's Guide to the Complex Terrain Dispersion Model Plus Algorithms for Unstable Situations (CTDMPLUS). U.S. EPA Report EPA/600/8-89/041.

[46] Lorimer, G. (1986). The AUSPLUME Gaussian-plume Dispersion Model. Australian EPA Report 86-02.

[47] Scire, J.S., Strimaitis, D.G., Yamartino, B.J. (2000). A User's Guide for the CALPUFF Dispersion Model (Version 5). Earth Tech Inc., Concord, Ma.

[48] Hurley, P. (2002). The Air Pollution Model (TAPM) Version 2: User Manual. *CSIRO Atmospheric Research Internal Paper n. 25*, CSIRO, Melbourne.

In: Ships and Shipbuilding
Editor: José A. Orosa

ISBN: 978-1-62618-787-0
© 2013 Nova Science Publishers, Inc.

Chapter 10

APPLYING THE LIFE CYCLE THINKING TO SEA PORTS: THE CASE OF A SLOVENIAN COMMERCIAL PORT

Boris Marzi[1], Stefano Zuin[2], Gregor Radonjič[3] and Klavdij Logožar[3]*

[1] Koper Port, Vojkovo nabrežje 38, 6000, Koper, Slovenia
[2] Venice Research Consortium, Venice, Italy
[3] University of Maribor, Faculty of Economics and Business, Maribor, Slovenia

ABSTRACT

Nowadays, sea ports are parts of the industrial network as constitute highly sophisticated area where logistics activities are concentrated due to e.g. distribution of different type of goods. For example, traffic demands changed the port role from a simple hinterland terminal to a complex logistic nodal point. However, port activities and operations may cause environmental impacts. Air pollution may result from ship traffic, ship accidents or land activities (road traffic), while water contamination are due to port maintenance, ship waste management, sewage treatment, etc. The port industry has then faced governmental mandates for achieving regulatory compliance, including environmental requirements. Meeting these requirements has been typically perceived as added costs.

This chapter introduces the Life Cycle Thinking (LCT) and how this approach may be applied to sea port activities and operations to transform port's environmental compliance into a business attribute that would produce an enhancement of a port's competitive position. In detail, the chapter highlights how the Life Cycle Assessment (LCA) methodology may be applied to port waste management and to transport routes inside the Koper port (case study). This study shows how Luka Koper company in practice endorse LCT, and to what extent, as a supportive tools into the business system of Luka Koper, in their attempt to become the leading port operator and global logistics solution provider serving the countries of Central and Eastern Europe.

Keywords: Life cycle assessment, carbon footprint, sea port, ship waste

---

* Corresponding Author address: Email:sz.cvr@vegapark.ve.it.

INTRODUCTION

The maritime transport is a catalyst of economic development and prosperity in different countries. Nowadays, sea ports play an important role for the global supply chain. Approximately 90% of Europe's international cargo trade and 40% of the intra-Community trade in goods passes through European ports. Also, more than 400 million passengers embark and disembark in European ports every year (European Commission, 2009). As consequence, the demand for increased capacity of marine transportation vessels, facilities, and infrastructure is expected to continue in the future. This demand is fuelled by a need to accommodate growing vessel operations for cargo handling activities and human population growth in coastal areas (U.S. Commission on Ocean Policy, 2004). As coastal areas continue to grow, there is a consequent increase in the demand for port services.

However, the expansion of port facilities, vessel operations, and commercial and recreational marinas may have adverse impacts on environment (Lam and Notteboom, 2012). The port industry is subject to closer scrutiny in terms of environmental regulatory compliance as significant changes in climate and their impacts are expected to become more pronounced in the next decades. Ships that call at ports are a major source of air pollutants such as CO_2, CO, SO_2, NOx, PM_{10}, $PM_{2.5}$, hydrocarbons and volatile organic compounds (VOC) (Lam and Notteboom, 2012). Coastal passenger shipping was found to be the dominate contributor of emissions in the passenger port of Piraeus in Greece (Tzannatos, 2010). As for Barcelona in Spain, the highest polluters are auto carriers among all other ship types (Villalba and Gemechu, 2011). Not only air emissions, but also water pollution and the effects on marine ecosystems are other major environmental concerns in Port of Rotterdam, as demonstrated by Ng and Song (2010). Other pollutants come from ballast water, fuel oil residue and waste disposal from ship operations as well as cargo residue. These marine pollutants are harmful to natural habitats located around port waters. In this context, the new EU Marine Strategy Framework Directive ensures that Member States are able to achieve "good environmental status" in marine waters covered by their sovereignty or jurisdiction by 2020. The growth of the marine transportation industry is accompanied by land-use changes, strengthening of the maritime access infrastructure, as well as over-water or in-water construction. The need for dredging operations and dredging disposal would lead to contaminated sludge from dredging that could cause environmental concerns (Peris-Mora et al., 2005). The management of port waste and noise pollution are other environmental priorities for the European port sector (ESPO, 2009).

In recent years, ports have made significant efforts to improve their environmental performance, as showed in the last ESPO Green Guide (ESPO, 2012), which revised and updated the Environmental Code of Practice that was produced by ESPO in 1994. Some ports have implemented the environmental management system (EMS) which is a systematic approach to manage a port's environmental programs for pollution prevention, protection and control (Florida and Davison, 2001). Several other ports are already planning for the development of manufacturing facilities to support a massive expansion of renewable energies (e.g. offshore wind power) (Rynikiewicz, 2011). The World Ports Climate Initiative (WPCI) has been set to initiate programs at ports that reduce greenhouse gas emissions and improve air quality. In other words, the port/public authorities have at their disposal different tools and/or approaches, e.g. self diagnosis method, air quality management by monitoring and measuring, established environmental standard (ISO 14001) or the EU Eco-Management and Audit Scheme (EMAS), needed to evaluate green port development to encourage green port development at various functional activities of port operations.

This chapter aims to evaluate how Life Cycle Assessment (LCA) may be used to identify possible improvements to ports management and services in the form of lower environmental impacts and reduced use of resources. The chapter will cover the analysis of environmental priorities of ports by means the LCA methodology, a cornerstone of the Life Cycle Thinking (LCT). The chapter will mainly focus on application of the LCA tool to port waste management and transport in Port of Koper (Luka Koper, Slovenia). A discussion on the research findings, policy implications and recommendations for the ports in general will be presented in the last section.

ENVIRONMENTAL PRIORITIES OF MARINE PORTS

The past years have seen increasing concerns of the environmental impact of port operations and development. Environmental considerations can be different for each port and depend on the specific location and the characteristics of the port area. In February 1996, ESPO commissioned the first environmental survey about ports in order to identify the issues which were significant for EU ports in the field of environment. This first environmental survey was carried out in 281 ports from 15 different European countries. In 2005 and 2009 the questionnaire has been updated to allow a comparison of the results of studies and also to investigate emerging trends in terms of environmental management and the progress that has been achieved over the years (ESPO, 2012).

According to these recent environmental surveys, the most prioritised environmental issues are: air quality management, energy conservation and climate change, noise management, waste and water management. The air emissions of CO, NOx, SO_2, hydrocarbons, VOC, and particulates matters by port operations are a high priority for ports because air pollution is often at the heart of the political and societal debate about port development projects. Ships that call at ports are a major source of air pollutants, but also landside activities, e.g. cargo operations at terminals, are other sources of airborne emission. Appropriate control mechanisms to manage and reduce port related air pollution are then needed for port authorities. Energy consumption and greenhouse gas emissions from shipping and the port sector are increasingly in the focus of public and political attention. In addition, ports are often also the location where industries use a lot of energy. Noise pollution is the problem most often cited in connection with transport. The main sources of noise in ports are the direct port operational activities, traffic (road, rail, ships) and industries based inside the ports. Ports are very often key hubs for passengers and also often the location where industrial activities take place. All these activities generate different type of waste. Hence improper waste control and treatment can be a great strain on the natural environment surrounding the port area. Ports have to handle these waste by a management plan, which regulates how the port is handling all kinds of waste. Finally, port activities may also affect water quality and marine ecosystems. Water pollution comes from ballast water, fuel oil residue and waste disposal from ship operations as well as cargo residue. An appropriate water management is then needed in ports.

The EC's action looks after the sustainable development of all ports in Europe, promoting industrial efficiency, reducing environmental impact and looking after working conditions and smooth integration of ports into the overall transport chain (EC, 2009). The environmental sustainability of ports, through e.g. the development of cleaner operation or the reduction of greenhouse gases, is a mandatory for the recent European ports' policy. The Adriatic Sea is an important maritime transport route used by merchant ships in international and national trade. A significant number of important industrial centres are located along the western Adriatic coast and several Mid-European – and in many cases landlocked – countries heavily depend on the Northern Adriatic ports for the import of fossil fuels - energy. These sea ports are an integral part of an ever stronger transport and logistics chain which links Central and Eastern Europe with the Far East. The five seaports located at the northern tip of Adriatic sea - Trieste, Venice, Ravenna, Rijeka and Koper - handle more than 100 million tonnes of water-borne cargo every year. The cargo consists mainly of general cargo, containers, cars, ores and minerals, fossil fuels, chemicals and others types of cargo. Over recent years development has been further stimulated by exceptional growth in cargo throughput, especially container freight, together with the largest vehicle distribution centre in the Mediterranean. Consequently, port authorities of the Adriatic sea have to be proactive to ensure a sustainable development of ports, by improving their environmental performance, according to the recently EU ports policy.

LIFE CYCLE THINKING

There are different public policies (e.g. Integrated Product Policy of the EU), market (e.g. green procurement), and financial (e.g. environmental risk) drivers promoting the use of LCT in the design of products and services (Maxwell et al., 2006). LCT is a concept that integrates existing consumption and production strategies, preventing a piece-meal approach (UNEP, 2004). This approach is aimed to identify possible improvements to goods and services in the form of lower environmental impacts and reduced use of resources across all life cycle stages of product or service, i.e. "cradle to grave". The life cycle of products includes the raw material extraction and conversion, product manufacturing, packaging and distribution, product use and/or consumption, until the end of life (re-use, recycling of materials, energy recovery and ultimate disposal). The scope of LCT is to minimize the negative impacts at each life cycle stage, avoiding the transfer of the problem from one life cycle stage to another, from one geographic area to another and from one environmental medium (air, water, soil) to another (UNEP, 2004). For example, saving energy during the use phase of a product, while not increasing the amount of material needed to provide it. The life cycle perspective is an important building block of the EU's policy on sustainable consumption and production (EU, 2010), and of the United Nations Environment Program (UNEP, 2004). The LCT is practically based on a toolbox which includes different methods and tools, such as the Life Cycle Assessment (LCA), one of the most thorough and reliable analytical tool for the assessment of potential impacts of a product or service system.

LCA is a systematic assessment of the potential environmental impacts of a defined good or service throughout all life-cycle stages, contributing to the production of the product under investigation (including auxiliary materials), its use and disposal (Hauschild, 2005). The assessment typically covers a broad range of environmental impacts, such as climate change, resource depletion, and toxicity on human health due to releases of e.g. chemicals. The International Organization for Standardization (ISO) defines LCA as the "compilation and evaluation of the inputs, outputs and the potential environmental impacts of a product system throughout its life cycle" (ISO, 2006a). According to ISO 14040 series standards (ISO 2006a, b), LCA consists in four main phases (fig. 1): (i) defining the goal and scope of the study, (ii) establishing a life cycle inventory which aggregates all inputs from and outputs to the environment within the system boundaries, (iii) performing a life cycle impact assessment which translates the inventory into potential impacts of the system on the environment and (iv) interpreting the results from the assessment to provide consistent support to decision-makers in relation to the goal and scope of the study. The figure 1 illustrates the life cycle assessment framework as described by ISO.

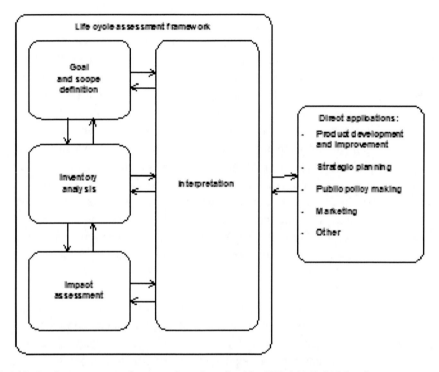

Figure 1. Life Cycle Assessment framework as described by ISO 14040:2006 series.

As shown in figure 1, LCA is an iterative process. The choices made during the course of the study can be modified when new information is available. The results of LCA are a set of environmental indicators (Global Warming Potential, Resource Depletion, Ozone Layer Depletion, Human Toxicity, etc.) normally presented on a chosen functional unit basis. The functional unit provides a reference to which the input and output flow data relate. Another important aspect of the scope statement is the definition of the initial boundaries of the system which is being analyzed, so that the input and output flows at the boundary can be clearly identified. Although LCA is the only tool that examines the environmental impacts of a product or service throughout its life cycle and it is also an ISO standardized methodology, LCA does not take into account spatial and temporal information (European Commission - Joint Research Centre, 2011). A significant number of LCA studies have already been conducted on several different subjects, e.g. clothing, polymer, building products, bottles, sanitary landfill, means of transport, etc.

CASE STUDY: LUKA KOPER

The Port of Koper is an international cargo port and was established in 1957. It lies at the northern edge of the Adriatic Sea, at its deepest indentation into the European continent, providing an accessible Mediterranean port with railway track and road links through inter-modal and combined transport. Luka Koper is involved in international trade and international operations. It is a multi-purpose port, equipped and prepared for handling and warehousing of all types of goods. The basic port activity is carried out at specialised terminals, which are technically and organisationally suitable for handling and warehousing of specific cargo groups: general cargo, fruit and light-perishable goods, livestock, containers, vehicles, timber, dry bulk and liquid cargoes. Today, the annual maritime throughput is over 15 million tonnes. In 1997 the Luka Koper company was awarded the International Standardization Organization (ISO) 9002 certificate for its quality services, followed by the ISO 9001. By controlling and systemically reducing the environmental impacts, investing in separate collection of waste and its processing, setting up sea monitoring and establishing co-operation with international organisations, Luka Koper systemically approached environmental management and adaptation to the standards of the European Union. In 2000 the company received the ISO 14001 environmental certificate, in 2007 the ISO 22000:2005 food safety system certificate, and in 2008 the BS OHSAS 18001:2007 occupational health and safety certificate. In 2009 the company verified that its environment management processes were congruent with the provisions of Eco-Management and Audit Scheme (EMAS), according to the regulation n. 761/2001 of the European parliament for voluntary participation by organisations in a Community EMAS. Also, Luka Koper has implemented a conservation strategy which dictated both a reduction in the consumption of natural resources as well as a reduction in the impact of the company's activities on the environment. This policy shall continue into the future too, ensuring operations compliant with the standards and expectations of business, the demands of legislation and the satisfaction of the broader community. Finally, the company will further consider the principles of sustainable development and responsible environment management, in order to respond to more complex environmental and economic concerns.

Life Cycle Assessment of Shipboard Waste Management

Ports generate, receive and handle a large volume of different kinds of waste (e.g. garbage, bilge waters, oil, sludge, etc.) every year and these wastes are viewed by ports as one of the most important environmental issues with which they are faced (Georgakellos, 2007). In detail, cruise ships generate large amounts of waste that requires disposal in a responsible manner. Whilst MARPOL 73/78 allows for disposal at sea of several types of treated waste once a vessel is 12 miles offshore (Butt, 2007), there is a need to encourage sustainable action for shipboard waste management, rather using the sea as a convenient dump. In order to achieve this there is a need for all ports to manage their waste in a sustainable way, e.g. by providing adequate recycling facilities in the port itself. In this context, the Environmental and Health Safety Department of Luka Koper decided to perform a LCA of management of shipboard waste (Zuin et al., 2009).

The overall goal of the study was to analyse shipboard waste streams and to assess the impacts deriving from their management in order to detect the critical stages. The functional unit was defined as the average annual amount of cargo-generated waste collected and managed in Luka Koper in year 2007, i.e. 2200 tonnes/year. The system boundaries included waste recovery from cargo vessels dock, transport, collection and final disposal. Cargo vessel-generated waste were considered in study, which are separated into solid waste (e.g. glass, plastic, metals), biodegradable waste (i.e. kitchen waste), wastewater (i.e. oily bilge waters), and other cargo residues. Gray and black wastewaters of ships were excluded as they are usually discharged into marine waters.

The management of shipboard waste in port of Koper includes the following processes: (i) a collection unit of both solid (1600 ton/year) and kitchen (100 ton/year) waste, (ii) an oil/water separator, where bilge waters (600 ton/year) are separated in oils and water by a physical process, (iii) and a wastewater treatment plant where all wastewaters (250 ton/year) are treated in a small plant with an average capacity size of 800 per capita equivalents. Finally collected solid waste is disposed off in a municipal landfill, while collected kitchen waste is incinerated in a waste-to-energy plant, according to the EU regulation 1774/2002 concerning the safely disposal of international catering waste from ports and airports (EC, 2002); on the contrary, separated oils are used as secondary fuel in a cement production plant. Also transport of waste by small barge at port, and by lorry inside and outside the port to a recovery facility and final disposal was included in the study. Data (e.g. quantities and composition of shipboard waste, average distance of transport, energy consumptions of different processes located inside the port) was provided by Luka Koper. Primary data was then integrated with secondary sources with regard to a cement production plant (van Oss and Padovani, 2002 ; 2003) and the ecoinvent v2.0 database included in the LCA software used (SimaPro 7.1).

The impact assessment of shipboard management was performed by applying the CML and Eco-Indicator '99 methods, as these methods are used to represent the midpoint (CML-method) and endpoint approaches (Eco-Indicator '99) (Goedkoop and Spriensma, 2000; Guinée et al., 2001).

With regard to the life cycle impact assessment, table 1 shows the results of ten impact categories obtained by CML 2 base line 2000 method. Absolute values (i.e. characterised indicator results) of each impact category are reported in table 1 for the two phases, waste collection phase and waste treatment phase. For almost all CML categories, the waste treatment phase has the greatest contribution.

For example, for the Eutrophication potential (EP) category, the waste treatment phase contributes with 3980 kg PO_4^{3-} equivalent (eq), while the waste collection phase contributes with 10.3 kg PO_4^{3-} eq. The most relevant contribution in EP category are the release of organic compounds (chemical oxygen demand ; COD), ammonium ions and nitrate from municipal landfill to water. Similarly, for the global warming potential (GWP) category, the waste treatment phase contributes with about 1000000 kg CO_2 eq, due to air release of biogenic methane and carbon dioxide from landfill.

The disposal of solid waste in landfill is also the most relevant contribution to the photochemical ozone creation potential (POCP), with 137 kg C_2H_4 eq, as result of biogenic methane released to air from landfill; for the marine aquatic, freshwater aquatic and terrestrial ecotoxicity potential (MAETP, FAETP and TETP) and human toxicity potential (HTP) categories, the waste treatment contributes with more than 90% in these categories. On the contrary, for the abiotic depletion potential (ADP), acidification potential (AP) and ozone layer depletion (ODP), the waste treatment phase presents benefits, with -5450 kg Sb eq, -1080 kg SO_2 eq, and -0.1 kg CFC11 eq, respectively. These negative values (i.e. environmental benefits) are due to te recovery of separated oil in cement plant. For example, for the ADP, the recovery of separated oil allows to avoid the extraction of oil crude (-5160 kg Sb eq) and hard coal (-45.6 kg Sb eq). For the AP, the benefits are due to -1050 kg of sulphur dioxide (SO_2) released to air, avoided emissions following oil recovery in waste treatment phase.

Impacts calculated with Eco-Indicator '99 are shown in table 2. The contribution of three Eco-Indicator'99 damage categories Human Health (HH), Ecosystem Quality (EQ) and Resources (R) is considered for the two phases considered. As showed in tab. 2, the results for HH damage category are greatly dominated by the waste treatment phase with 1.54 DALY; this is mainly due to carcinogenic substances (especially cadmium and arsenic) and greenhouses gases emissions (methane and carbon dioxide) associated with landfill of waste treatment phase. Also for the EQ damage category, the waste treatment is the most relevant phase with 3.45E+05 PDF*m^2*yr; the waste collection phase contributes with 1.13E+03 PDF*m_2*yr. The damage to EQ category is strongly related to release of cadmium, copper, lead, nickel and zinc from landfill to water, and to land occupation (dump site and traffic road) and land transformation (conversion of agricultural land to dump site). With regard to R damage category, the negative value (-1.66E+06 MJ) for waste treatment phase highlighting reduced impacts in comparison with waste collection phase. The expected increase of produced energy per kg of fossil fuels (resource) used is reduced for waste treatment phase. On the other words, the damage to R is in part reduced by oil recovery in cement plant.

The normalised results obtained considering the CML West Europe factors are reported in fig. 2. Although the normalization of characterized results is not an mandatory step in LCA methodology, it is strongly recommended to understand the relative importance and magnitude of the results for process (Guinée et al., 2001). In our case we have chosen a normalisation method applicable to the regional scale, where the different characterised impact scores are related to a common reference, such as the impact score for the Europe for year. Among the ones included in CML 2 method, the regional scale (West Europe) factors for normalisation have been chosen as they are the ones that best fit our study (Guinée et al., 2001).

According to this normalization, the most relevant categories to the total impacts are FAETP and MAETP (fig. 2). The disposal of solid waste in landfill is the prevalent process in both categories. This is due to heavy metals (copper, nickel and zinc) released in water from landfill. The benefits due to oil recovery in cement plant are also pointed out in ADP category.

Also the results of characterization obtained with Eco-indicator '99 method were normalized and then weighted to compare the environmental impacts on a same scale and better understand the magnitude of impacts. The obtained results, expressed as weighted final score (i.e. Eco-Indicator Point; Pt) for the three damage categories are showed in fig. 3.

According to this normalization, the most damaged categories are HH (4.09E+04 Pt) and EQ (2.70E+04 Pt) (fig. 3). For the R damage category, the benefits of oil recovery are quantified in -3.94E+04 Pt. As for the impact category ADP of the CML method, the recovery of separated oils as secondary fuels during the waste treatment phase allows to reduce in part the impacts by reducing the resources depletion. The analysis also indicates that approximately 95.3% of all environmental impacts are caused by the final treatment of shipboard waste, and the landfill disposal of solid waste contributes the most to the environmental load. Landfill is responsible for over two third of contributions to carcinogenic and ecotoxicity effects. With respect to waste collection phase, the contribution to damage categories is insignificant.

Table 1. Values of the characterization of environmental impacts according to CML methods for management of 2200 tonnes of shipboard waste per year

Impact category	Unit	Total	Waste collection	Waste treatment
ADP	kg Sb eq	-5.28E+03	1.67E+02	-5.45E+03
AP	kg SO_2 eq	-7.38E+02	3.42E+02	-1.08E+03
EP	kg PO_4^{3-} eq	3.99E+03	1.03E+01	3.98E+03
GWP100	kg CO_2 eq	1.04E+06	2.42E+04	1.02E+06
ODP	kg CFC-11 eq	1.03E-01	1.19E-03	-1.05E-01
HTP	kg 1,4-DB eq	2.18E+05	5.57E+03	2.12E+05
FAETP	kg 1,4-DB eq	3.10E+06	2.98E+03	3.10E+06
MAETP	kg 1,4-DB eq	1.07E+09	7.46E+06	1.07E+09
TETP	kg 1,4-DB eq	2.07E+03	1.14E+02	1.95E+03
POCP	kg C_2H_4	1.50E+02	1.31E+01	1.37E+02

Legend: abiotic depletion potential (ADP); acidification potential (AP); eutrophication potential (EP); global warming potential for the time horizon of 100 years (GWP100); ozone layer depletion potential (ODP); human toxicity potential (HTP); marine aquatic, freshwater aquatic and terrestrial ecotoxicity potential (MAETP, FAETP and TETP); photochemical ozone creation potential (POCP)

Table 2. Values of the damage category according to Eco-indicator 99 method (Hierarchist perspective) for management of 2200 tonnes of shipboard waste per year

Damage category	Unit	Total	Waste collection	Waste treatment
HH	DALY	1.57E+00	3.45E-02	1.54E+00
EQ	PDF*m^2*yr	3.46E+05	1.13E+03	3.45E+05
R	MJ surplus	-1.64E+06	1.56E+04	-1.66E+06

Legend: HH = Human Health; EQ = Ecosystem Quality; R = Resources; DALY = Disability Adjusted Life Years; PAF = Potentially Affected Fraction; PDF = Potentially Disappeared Fraction

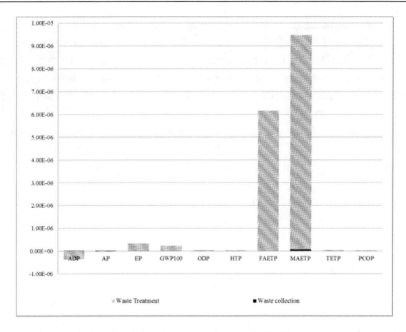

Figure 2. Normalized impact values applying the CML2001 Method (West Europe, 1995). The normalization is applied on ten category indicators selected for the characterisation on impacts: abiotic depletion potential (ADP); acidification potential (AP); eutrophication potential (EP); global warming potential for the time horizon of 100 years (GWP100); ozone layer depletion potential (ODP); human toxicity potential (HTP); marine aquatic, freshwater aquatic and terrestrial ecotoxicity potential (MAETP, FAETP and TETP); photochemical ozone creation potential (POCP).

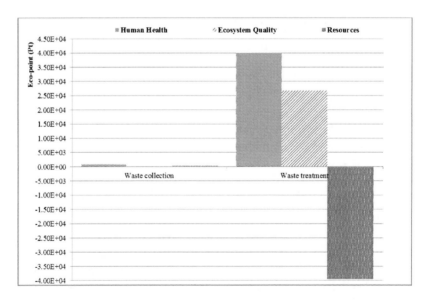

Figure 3. Normalized and weighted damage values applying the Eco-Indicator '99 (H/A) method. The normalization is applied on three damage categories: Human Health, Ecosystem Quality and Resources.

Greenhouse Gas Inventory of Activities and Transport Inside the Port Basin

The most important greenhouse gas (GHG) that derives from human activity is CO_2. The carbon footprint derives from an equation that enables to calculate the quantity of greenhouse gases expressed as equivalents of CO_2 for a certain activity. Luka Koper participated in the EU project that encourages the Mediterranean ports to decrease greenhouse emissions in the air and at the same time encourages the efficient use of energy. Carbon footprint quantification, analysis and reduction are key to preventing climate change, by enhancing energy efficiency, mitigating carbon emissions by means of green energy and then compensating for remaining GHG emissions by investing in carbon offsets, with the final goal of becoming carbon neutral. Luka Koper participated in all phases of the project: preparing a port greenhouse emission report, calculating the port CO_2 footprint, and benchmarking the best practices in order to reduce the emissions of greenhouse gases and to develop port specific environmental action plan towards the climate change mitigation.

Reducing greenhouse emissions and energy saving are some of objectives that Luka Koper has already implemented in its environmental policy and strategy. Also, the port of Koper has already introduced modern and energy-efficient technologies (e.g. reduction of consumption of natural resources, reduction of detrimental impacts to the environment), while different approaches for the reduction of releases to the environment are ongoing in Luka Koper to improve its environment management system.

In order to quantify the carbon footprint of port activities and transport inside the port, a detailed greenhouse gas inventory was performed. Compiling a greenhouse gas inventory is a step-by-step process. The inventory has included: data collection, estimation of emissions and removals/reduction processes, checking and verification, uncertainty assessment and reporting. For the calculation of the carbon footprint the following steps were performed according to the ISO 14064:2006 series (ISO, 2006c; 2006d) and guideline of the Intergovernmental Panel on Climate Change (IPCC; 2006): (i) establishing operational boundaries (i.e. identifying GHG emissions and removals, categorizing GHG emissions and removals into direct emissions, energy indirect emissions and other indirect emissions) (ISO 2006c; 2006d); (ii) quantification of GHG emissions and removals (i.e. identification of GHG sources and sinks - GHG inventory, selection of quantification methodology, collection of GHG activity data, selection or development of GHG emission or removal factors, calculation of GHG emissions and removals) (ISO 2006c; 2006d; IPCC, 2006).

The carbon footprint as calculated for the Port of Koper in year 2011 amounted to 53,273 tons of CO_2 eq or 3.39 kg CO_2 eq for each handled ton of cargo, where in year 2010 there were 49013 tons CO_2 eq directly or indirectly produced (i.e. 3.53 kg CO_2 eq for each handled ton of cargo). All the direct and indirect port activities that produce CO_2 emissions were considered and included in the calculations, as well as vessels' emissions over time when vessels are moored in the port and which represents 40% of all CO_2, CH_4 and N_2O emissions.

In 2012 the Port of Koper produced 51,520 tons of CO_2 eq, which is equivalent to 3.20 kg of CO_2 eq per ton of cargo handled.

Calculating the carbon footprint is only the beginning of the battle against the sources of greenhouse gases. Without taking the necessary measures to reduce emissions, this battle is worthless. The goal of Port of Koper is to develop balanced and harmonized port operations in order to improve the weak points concerning greenhouse emissions of the ports. As collection and handling of port waste are resource-intensive processes, the port of Koper have ascertained the importance to optimize the performed GHG inventory, and decided to compile a GHG inventory for the waste collection and handling processes inside the port as well. The carbon footprint has thus become an important environmental tool which can be used as an environmental performance indicator and a powerful strategic tool for sustainable investments.

CONCLUSION

During the last 10 years, sea ports have improved their environmental performance. The increasing trend for ports to perform an environmental policy, to publish an annual environmental report, or to establish activities and procedures to manage their waste, are examples of proactive actions regarding sustainability. But sustainability requires a greater focus on integrating methods, tools, and approaches including LCT. This study showed how the life cycle perspective may be helpful to support the port environmental management. LCA tool may be used as a basis for decision support processes within the planning of port activities as it allows to identify the contribution of different port operations, providing useful and helpful information to all stakeholders involved, i.e. port authorities, local municipalities, etc. It should be emphasized that the port authorities are not necessarily and directly responsible for the activities, products and services of the components of the logistic chain; but their overarching administrative role and presence is leading them as the main player for any environment related issues in the whole port area. For example, within the waste management, the port authorities could act as facilitators of procedures and of communication between the different parties involved in the waste management. In addition, ports cannot act alone while carrying out any green ports strategies; they need collaborations from stakeholders including public policy makers, market players and community, because most strategies are not unilateral.

Ports are often located in an urban area. They influence the whole territory from different points of view: economic, environmental and mobility systems. In this context it is needed to balance ports and urban community exigencies in a way much more critical compared to the past. We think that cost-benefit analysis and the environmental impact assessment are useful tools for identifying, predicting, evaluating and mitigating the biophysical, social, and other relevant effects of development proposals for ports planning.

The company Luka Koper has shown a great interest in the application of LCT, through the investigation of impacts posed by shipboard waste management, and the quantification of carbon footprint of port activities and transport. This case study should be a moment of comparison between the different actors, such as port users, community groups, business community, local government, etc., to emphasize the importance of LCT as a potential perspective for port development. The examples of good practices of environmental sustainability in Luka Koper may be suggested as action to be exported to other ports, in order to ensure the development of port compatible with the needs of environment and conservation of resources.

REFERENCES

Butt, N., (2007). The impact of cruise ship generated waste on home ports and ports of call: A study of Southampton. *Marine Policy 31*, 591-598.

Eggleston H.S., Buendia L., Miwa K., Ngara T., Tanabe K., (2006). IPCC Guidelines for National Greenhouse Gas Inventories.

EC - European Commission - Joint Research Centre - Institute for Environment and Sustainability, (2011). International Reference Life Cycle Data System (ILCD) Handbook- Recommendations for Life Cycle Impact Assessment in the European context. First edition November 2011. Luxemburg. Publications Office of the European Union.

EC - European Commission, (2002). Regulation No 1774/2002 of the European Parliament and of the Council of 3 October 2002 laying down health rules concerning animal by-products not intended for human consumption. *Official Journal of the European Communities L 273*, 1-95.

EC -European Commission. (2009). Communication from the Commission to the European Parlament, the Council, the European Econimc and Social Committee and the Committee of the Regions. Strategic goals and recommendations for the EU's maritime transport policy until 2018. Brussels, 21.1.2009, COM(2009) 8 final.

ESPO - European Sea Port Organization, (2012). Green Guide. Towards excellence in port environmental management and sustainability.

ESPO - European Sea Ports Organisation, (2009). ESPO / EcoPorts Port Environmental Review. European Sea Ports Organisation's Review of Environmental Benchmark Performance in collaboration with the EcoPorts Foundation (EPF).

EU - European Union, (2010). Making sustainable consumption and production a reality. A guide for business and policy makers to Life Cycle Thinking and Assessment. Luxembourg: Publications Office of the European Union, 2010.

Florida, R., Davison, D., (2001). Gaining from green management: Environmental management systems inside and outside the factory. *California Management Review* 43, 64-85.

Georgakellos, D.A., (2007). The use of the deposit-refund framework in port reception facilities charging systems. *Marine Pollution Bulletin 54*, 508-520.

Goedkoop, M., Spriensma, R., 2000. The Eco-indicator 99: A damage oriented method for life cycle impact assessment. PRé Consultants, Amersfoort, Netherlands, 2000.

Guinée, J. B, Gorrée, M., Heijungs, R., Huppes, G., Kleijn, R., de Koning, A., (2001). Life cycle assessment: an Operational Guide to the ISO Standards - Institute of Environmental Sciences, 2001.

Hauschild, M.Z., (2005). Assessing environmental impacts in a life-cycle perspective. *Environ. Sci. Technol. 39*, 81A–88A

ISO - International Organization for Standardization, (2006a). Environmental management - life-cycle assessment - principles and framework. ISO 14040:2006.

ISO - International Organization for Standardization, (2006b). Environmental management – life cycle assessment - requirements and guidelines. ISO 14044:2006.

ISO - International Organization for Standardization, (2006c). Greenhouse gases - Part 1: Specification with guidance at the organization level for quantification and reporting of greenhouse gas emissions and removals. ISO 14064-1:2006.

ISO - International Organization for Standardization, (2006d). Greenhouse gases - Part 2: Specification with guidance at the project level for quantification, monitoring and reporting of greenhouse gas emission reductions or removal enhancements. ISO 14064-2 :2006.

Lam, J.S.L., Notteboom, T. (2012). The green port toolbox: A comparison of port management tools used by leading ports in Asia and Europe. Paper presented at International Association of Maritime Economists (IAME) Conference, Taipei, Taiwan, 5-8 September 2012.

Maxwell, D., Sheate, W., van der Vorst , R., (2006). Functional and systems aspects of the sustainable product and service development approach for industry. *J. Cleaner Production 14*, 1466-1479.

Ng, A.K.Y., Song, S., (2010). The environmental impacts of pollutants generated by routine shipping operations on ports. *Ocean & Coastal Management* 53, 301-311.

Peris-Mora, E., Diez Orejas, J.M., Subirats, A., Ibanez, S., Alvarez, P., (2005). Development of a system of indicators for sustainable port management. *Marine Pollution Bulletin* 50, 1649–1660.

Rynikiewicz, C., (2011). European port cities as gateways to a green economy? *Network Industries Quarterly* 13, 23-25.

IPPC - The Intergovernmental Panel on Climate Change, (2006). Guidelines for National Greenhouse Gas Inventories. Prepared by the National Greenhouse Gas Inventories Programme, Eggleston H.S., Buendia L., Miwa K., Ngara T. and Tanabe K. (eds).

Tzannatos, E., (2010). Ship emissions and their externalities for the port of Piraeus – Greece. *Atmospheric Environment* 44, 400-407.

U.S. Commission on Ocean Policy. (2004). An Ocean Blueprint for the 21st Century. Final Report. Washington, DC, 2004.

UNEP - United Nations Environment Programme, (2004). Why take a Life Cycle Approach. UNEP. Division of Technology, Industry and Economics. Production and Consumption Branch, Paris, France. United Nations Publication.

van Oss, H.G., Padovani, A.C., (2002). Cement Manufacture and the Environment. Part I: Chemistry and Technology. *Journal of Industrial Ecology 6*, 89-105

van Oss, H.G., Padovani, A.C., (2003). Cement Manufacture and the Environment. Part II: Environmental Challenges and Opportunities. *Journal of Industrial Ecology 7*, 93-126.

Villalba, G., Gemechu, E.D., (2011). Estimating GHG emissions of marine ports - the case of

Zuin, S., Radonjič, G., Logožar K., Belac, E., Marzi, B., (2009). Life cycle assessment of ship-generated waste management of Luka Koper. *Waste Manag. 29*, 3036-3046.

INDEX

#

21st century, 179

A

abatement, 176
access, 202, 206
accommodation, vii
acid, 176, 188
adaptation, 210
adhesives, 157
adjustment, 11, 65, 72, 77
ADP, 212, 213, 214
aerosols, 175
aesthetic, 30
aesthetics, 30
air emissions, 182, 206, 207
air pollutants, 179, 188, 206, 207
air quality, 105, 168, 169, 171, 174, 175, 177, 181, 187, 188, 189, 192, 193, 196, 197, 199, 200, 201, 206, 207
air quality model, 192
airports, 87, 211
airways, 87
Alaska, 201
algorithm, 6, 12, 13, 15, 18, 20, 21, 33, 49, 66, 67, 178
ambient air, 176
ammonium, 211
amplitude, 73, 83, 91, 92, 93, 94
analytic expression, 18, 19, 43
annealing, 83
aquaculture, 126
architect, 18, 43
aromatic hydrocarbons, 175

arsenic, 212
artificial intelligence, 64, 81
Artificial Neural Networks, 64, 66, 68, 78
asbestos, 156, 163
Asia, 61, 82, 164, 218
Asian countries, 163, 164
assessment, 115, 119, 162, 169, 182, 187, 188, 189, 190, 191, 192, 193, 194, 195, 196, 198, 199, 200, 201, 205, 208, 211, 215, 216, 217, 218
atmosphere, 106, 111, 115, 116, 118, 157, 167, 170, 175, 179, 188, 193
authorities, 111, 125, 194, 206, 207, 208, 216
authority, 161
awareness, 86, 99

B

ban, 160
Bangladesh, 158, 159, 163, 164
base, 100, 127, 178, 188, 211
batteries, 157
benchmarking, 215
bending, 135, 144
benefits, 106, 123, 172, 181, 212, 213
Bézier curves, 11, 13, 21, 43
bias, 178
birds, 86
bleeding, 52
blogs, 61
boat, 40, 41, 198
boilers, 47, 48, 50, 51, 56, 57, 58, 59, 61, 191
bottom-up, 172, 178, 182
breathing, 121, 122
brothers, 86
buyer, 164
by-products, 217

C

cables, 125, 126, 127, 133, 134, 157
CAD, 11, 20, 39, 44
cadmium, 212
caesarean section, 165
CALPUFF model, 195, 196, 202
carbides, 175
carbon, 109, 170, 171, 173, 175, 176, 177, 178, 181, 182, 183, 188, 205, 211, 212, 215, 216
carbon dioxide, 176, 182, 211, 212
carbon emissions, 171, 177, 178, 181, 215
carbon neutral, 215
cargoes, 210
case studies, vii
case study, 205, 216
cash, 99
cash flow, 86
catalyst, 206
category a, 211, 212, 213
cation, 100
CCA, 82
centre of buoyancy, 9, 10, 11, 14
certificate, 162, 163, 164, 174, 210
certification, 168
challenges, 108, 155
chemical(s), 58, 109, 110, 116, 117, 157, 169, 171, 173, 174, 175, 177, 179, 180, 195, 208, 209, 211
chemical characteristics, 175
chemical properties, 173
China, 111, 114, 182
chronic diseases, 159, 163
cities, 177, 218
clarity, 7, 13
classes, 189
classification, vii, 66, 100, 115, 123, 128, 150, 169, 190, 198
cleaning, 53, 109, 157, 171, 174
climate, 169, 172, 174, 177, 179, 181, 182, 188, 206, 207, 209, 215
climate change, 169, 177, 207, 209, 215
closure, 53
clothing, 210
CO2, 105, 106, 167, 170, 176, 177, 182, 206, 211, 213, 215
coal, 48, 61, 174, 193, 194, 195, 212
coatings, 126, 157
collaboration, 156, 217
collisions, 155
combustion, 116, 167, 174, 175, 177, 178, 188, 189, 194
commercial, 8, 161, 172, 175, 176, 181, 182, 206
communication, 90, 216

community(ies), 164, 188, 210, 216
competition, 163
competitiveness, 177
compilation, 209
complement, 109
complexity, 8, 22, 65, 127, 156, 176, 192
compliance, 124, 187, 188, 196, 205, 206
composition, 89, 120, 127, 169, 174, 175, 177, 178, 179, 180, 189, 211
compounds, 174, 175, 211
compression, 116, 143, 144
computational fluid dynamics, 7
computer, 44, 66
computing, 21
concept map, 99
condensation, 52, 55, 172
conditioning, 53
conference, 156
configuration, 20, 79, 198
consensus, 83, 160, 177
conservation, 207, 210, 216
constituents, 175
construction, 11, 12, 19, 22, 42, 43, 61, 106, 125, 126, 127, 128, 129, 130, 131, 132, 140, 155, 159, 162, 178, 206
consumers, 58
consumption, 47, 48, 49, 60, 105, 106, 169, 172, 176, 177, 178, 182, 189, 190, 191, 194, 202, 207, 208, 210, 215, 217
consumption rates, 191
containers, 106, 109, 115, 117, 118, 121, 157, 189, 208, 210
contamination, 176, 205
contour, 27, 197, 199
controversies, 177
convention, 93, 155, 161
cooling, 57, 157, 175
copper, 212
corrosion, 117, 118, 127, 134, 136, 139, 140, 142, 144, 145, 146, 147, 148, 149
cost, 6, 47, 48, 49, 106, 111, 172, 173, 176, 177, 182, 216
cost-benefit analysis, 216
covering, 168
cracks, 119
critical analysis, 177
critical period, 73
crown(s), 130, 139, 140
crude oil, 48, 181
CSA, 78
customers, 123
cycles, 61, 169

Index

D

damages, 10, 114, 120
damping, 76
data collection, 215
data processing, 187, 188
database, 7, 8, 211
deaths, 115, 164
decay, 76
decoding, 87
decomposition, 21
decontamination, 164, 165
defects, 145, 147, 148
deficiencies, 110, 159
deformation, 127, 135, 136, 137, 138, 141
degradation, 145, 146, 149, 150, 159, 169, 177
degradation mechanism, 146, 150
delegates, 156
Denmark, 61, 181, 202
Department of Energy, 47, 105, 125, 155, 167
deposits, 109
depth, vii, 35
derivatives, 8, 12, 13, 14, 25, 31, 32
designers, 8, 9, 123
destruction, 116, 120, 122, 171
detection, 149, 176
developed countries, 163, 164
developing countries, 176
deviation, 23, 57, 99, 110
diesel engines, 48, 49, 60, 170, 173, 177, 191, 194
diesel fuel, 175
diffusion, 193, 202
directionality, 93
disaster, 111, 149
discharges, 168
discs, 119
diseases, 159, 163, 188
dispersion, 121, 177, 187, 188, 189, 192, 193, 196, 198, 202
displacement, 6, 8, 9, 27, 42, 70, 71, 72, 80, 81, 159
disposition, 51
distilled water, 57
distribution, 8, 9, 10, 11, 29, 38, 42, 43, 137, 172, 178, 193, 195, 198, 205, 208
DME, 88, 89, 90, 95, 96
dosing, 58
draft, 10, 11, 18, 23, 35, 161
draught, 7, 34
drawing, 127, 201
drinking water, 99

E

Eastern Europe, 205, 208
economic development, 206
economic values, 115
economics, 106
ecosystem, 172, 188
education, 99
educational objective, 86
election, vii
electricity, 190, 191, 197
electromagnetic, 88, 145, 146, 148, 149, 150
electromagnetic fields, 110
electromagnetic waves, 88
elongation, 143
emergency, 51, 56, 58, 157, 162, 170
emergency preparedness, 162
emergency response, 157
emission, 106, 167, 169, 170, 171, 172, 173, 175, 176, 177, 178, 179, 180, 181, 182, 187, 188, 189, 190, 191, 192, 193, 194, 195, 196, 198, 200, 202, 207, 215, 218
emitters, 175, 196
employers, 125
employment, 156
EMS, 206
energy, 47, 48, 49, 116, 117, 118, 121, 126, 167, 179, 180, 183, 189, 207, 208, 211, 212, 215
energy conservation, 207
energy consumption, 49, 189, 211
energy efficiency, 48, 215
energy recovery, 208
enforcement, 162, 168
engineering, vii, 63
enlargement, 193, 198, 200
environment(s), 108, 110, 121, 141, 143, 156, 160, 161, 162, 163, 164, 167, 168, 171, 177, 180, 188, 201, 202, 206, 207, 209, 210, 215, 216
environmental degradation, 159, 177
environmental effects, 48
environmental impact, vii, 60, 105, 106, 164, 167, 179, 188, 201, 205, 207, 208, 209, 210, 212, 213, 216, 217, 218
environmental issues, 177, 207, 210
environmental management, 206, 207, 210, 216, 217
environmental policy, 215, 216
environmental protection, 163
Environmental Protection Agency (EPA), 192, 198, 202, 203
environmental sustainability, 208, 216
EPC, 173
equilibrium, 9

equipment, 48, 55, 61, 88, 106, 110, 115, 117, 120, 131, 146, 147, 156, 157, 159, 169, 170, 174
erosion, 57, 118
Europe, 27, 164, 165, 168, 200, 205, 206, 208, 212, 214, 218
European Commission, 156, 164, 206, 209, 217
European Community, 156, 202
European Parliament, 105, 165, 166, 201, 217
European policy, 169
European Union (EU), 160, 162, 164, 201, 206, 207, 208, 210, 211, 215, 217
evacuation, 55, 56
evidence, 177
examinations, 134
expert systems, 100
exports, 176, 181
external shocks, 118
externalities, 179, 201, 218
extraction, 57, 165, 208, 212

F

fabrication, 101
facilitators, 216
fiber, 132
filters, 53, 178
financial, 164, 208
fires, 87, 111, 192
fishing, 5, 7, 44, 198
fixation, 48
flame, 52
flatness, 67
flaws, 146, 147
flexibility, 6, 27, 43, 108, 127, 132
flight(s), 86, 87, 177
flotation, 10
fluid, 7, 109, 118, 165, 191
food safety, 210
force, 111, 126, 156, 160, 162, 163, 168, 173, 174, 179
forecasting, 123, 202
formaldehyde, 182
formation, 119
formula, 65
fractures, 109, 140
fragments, 115, 119
France, 202, 218
frequency distribution, 198
freshwater, 212, 213, 214
friction, 135
fuel consumption, 47, 60, 105, 106, 172, 176, 177, 178, 182, 189, 190, 191, 194
fuel prices, 48, 173

fuzzy sets, 100

G

garbage, 210
geometry, 7, 22, 89
Germany, 44
GHG, 215, 216, 218
gland, 55, 56, 57
global scale, 187
global warming, 211, 213, 214
goods and services, 208
governments, 163
gravity, 8, 9, 10, 13, 16, 17, 31, 32
Greece, 179, 201, 206, 218
green management, 217
greenhouse(es), 167, 188, 212
greenhouse gas (GHG), 176, 206, 207, 208, 215, 218
greenhouse gas emissions, 176, 206, 207, 218
grouping, 143
growth, 48, 146, 177, 206, 208
guidance, 87, 218
guidelines, 99, 156, 162, 217

H

habitats, 179, 206
halogen, 174
harmful effects, 160, 161
hazardous materials, 161, 162
hazardous substances, 156, 158, 160, 162
hazardous waste(s), 156, 160, 163, 164, 165, 166
hazards, 108, 115, 122, 157, 163
health, vii, 156, 160, 161, 162, 163, 169, 171, 172, 175, 178, 179, 188, 209, 210, 217
health effects, 175, 178, 188
heat transfer, 59
heavy metals, 156, 157, 161, 174, 212
height, 10, 23, 29, 30, 33, 34, 35, 59, 89, 193, 195, 198
hemp, 128
heterogeneity, 178, 192
historical data, 120
history, 82
homes, 111
Hong Kong, 155, 156, 161, 162, 164, 165, 166
housing, 109
human, 6, 68, 86, 108, 109, 110, 114, 123, 146, 150, 156, 161, 162, 167, 169, 188, 206, 209, 212, 213, 214, 215, 217
human activity, 215
human brain, 68

Index

human health, 156, 161, 162, 169, 188, 209
hybrid, 15, 18, 33, 79, 178, 183
hydrocarbons, 119, 156, 161, 165, 175, 191, 206, 207
hypothesis, 171
hysteresis, 72, 73, 75

I

ICS, 163
ID, 95
ideal, 77, 106
identification, 81, 178, 182, 188, 215
ignition source, 116
image, 100, 116
IMO, v, vi, vii, 105, 106, 155, 156, 159, 161, 162, 163, 167, 168, 173, 174, 176, 177, 179
impact assessment, vi, 119, 187, 188, 189, 192, 193, 194, 196, 199, 200, 201, 209, 211, 216, 217
imports, 176, 181
improvements, 164, 175, 177, 207, 208
indentation, 135, 210
India, 159, 163, 164
industrial sectors, 188
industry(ies), 65, 105, 106, 109, 115, 122, 123, 125, 149, 155, 159, 205, 206, 207, 218
infrastructure, 107, 156, 206
injury, 114
inspections, 111, 134, 145, 146, 149
insulation, 157
integration, 13, 17, 31, 208
integrity, 105, 106, 107, 108, 134
interaction effect, 195
interaction effects, 195
interface, 202
interference, 52, 147
International Maritime Organization (IMO), v, vii, 105, 123, 153, 155, 167, 168, 172
international standards, 168
international trade, 210
interpolating B-spline, 20
interrogations, 95
intron, 150
investment, 48, 100, 111, 164
investments, 48, 216
ions, 175, 211
irrigation, 81, 109
isolation, 53
issues, 108, 124, 156, 176, 177, 178, 188, 207, 210, 216
Italy, 159, 187, 193, 198, 201, 205
iteration, 45

J

Japan, 82, 159
Jordan, 25
jurisdiction, 160, 161, 170, 174, 206
justification, 139, 140, 143

K

K^+, 67
knots, 39
Korea, 159
Kyoto Protocol, 176

L

Lagrange multipliers, 67
lasers, 99
Latin America, 81, 83
laws, 115
lead, 110, 118, 134, 139, 143, 157, 196, 206, 212
leakage, 108, 146
leaks, 108, 124
learning, 79, 100
legislation, 160, 210
life cycle, 208, 209, 211, 216, 217
lifetime, 101, 171, 181
light, 48, 93, 157, 175, 210
linear function, 78
linear systems, 81
liquefied natural gas, 47, 106
liquid phase, 56
liquids, 109, 115, 118, 157
livestock, 210
local government, 216
logistics, 205, 208
low temperatures, 108, 109
LTD, 165
lubricants, 157, 159
Luxemburg, 217

M

machine learning, 82
machinery, vii, 48, 120, 133, 157
magnetic field, 146, 148
magnetic properties, 125, 146
magnets, 147, 148
magnitude, 111, 120, 188, 196, 212
majority, 64, 131, 140
man, 128

management, 7, 160, 162, 177, 205, 206, 207, 210, 211, 213, 215, 216, 217, 218
manipulation, 6, 8
manufacturing, 111, 206, 208
mapping, 7, 66
marginal costs, 172
marine environment, 168
mass, 9, 90, 119, 146, 170, 173, 175, 176, 177, 188, 190, 191, 192, 193
material handling, 191
materials, 110, 118, 126, 157, 160, 161, 162, 188, 189, 193, 194, 208
mathematical programming, 100
matrix, 14, 17, 25, 28, 31, 32, 34, 39, 67, 96, 97, 168
matter, 156, 170, 172, 173, 175, 178, 181, 188, 201
maximum of the curve (XMAX), 10, 11, 13, 40
measurement(s), 145, 170, 173, 175, 178, 179, 181
media, 114
Mediterranean, 169, 179, 200, 208, 210, 215
membership, 160
mercury, 156, 157
messages, 87, 178
meta-analysis, 172
metals, 156, 157, 161, 174, 175, 211, 212
methanol, 111
methodology, 7, 9, 36, 64, 66, 81, 115, 169, 176, 189, 190, 194, 205, 207, 209, 212, 215
Mexico, 110, 112
microstructure, 175
military, 85, 86, 87
missions, 105, 106, 167, 172, 173, 175, 176, 178, 180, 182, 215, 218
mixing, 58, 181, 193, 195, 198
modelling, 170, 172, 176, 177, 178, 180, 182, 187, 188, 192, 193, 195, 196, 198, 202
models, vii, 64, 65, 86, 96, 99, 121, 171, 177, 182, 187, 189, 192, 193, 202
modifications, vii, 8
modules, 8, 9, 195
moisture, 141
molecular mass, 175
morphology, 175
mortality, 163
motivation, 43

N

natural disaster, 111
natural gas, 47, 62, 106
natural habitats, 206
natural resources, 210, 215
navigation system, 85, 86, 87, 101
Netherlands, 217

neural networks, 78, 79, 81
neurons, 68, 69, 78
neutral, 215
New Zealand, 176, 181
next generation, 61
NGOs, 156
nickel, 212
nitrogen dioxide, 170
nodes, 195, 198
non-OECD, 160
North America, 172, 180, 200
Norway, 61, 159, 180
nucleation, 119
nuclei, 172, 195

O

occupational health, 210
oil, 47, 48, 49, 52, 53, 60, 61, 110, 111, 117, 123, 126, 157, 165, 168, 170, 171, 172, 174, 175, 178, 180, 181, 190, 191, 195, 206, 207, 208, 210, 211, 212, 213
olefins, 111
operating costs, 60, 61
operating range, 64, 81
operating system, 194
operations, 53, 106, 109, 125, 128, 157, 162, 171, 173, 187, 188, 191, 192, 194, 195, 205, 206, 207, 210, 215, 216, 218
optical properties, 181
optimization, 7, 67, 78, 83, 86, 99
ores, 208
organ, 160
organic compounds, 173, 206, 211
Organization for Economic Cooperation and Development (OECD), 160, 164
oscillation, 73, 74, 75, 76
oxygen, 58, 108, 211
ozone, 171, 179, 180, 188, 200
ozone layer, 156, 169, 174, 212, 213, 214

P

Pacific, 61, 82
Pakistan, 159, 163, 164
parallel, 12, 37, 72, 128, 132, 133, 159
Parliament, 105, 165, 166, 201, 217
particle mass, 177
partition, 101
PCBs, 157, 188
PCDD/Fs, 188
performance indicator, 216

permission, 11
petroleum, 111, 167, 174
plant type, 48, 61
plants, 48, 49, 61, 82, 109, 110
platform, 27
pleasure, 189
PM, 170, 175, 177, 188, 191, 194, 196
polar, 182
policy, 146, 149, 169, 172, 176, 177, 182, 207, 208, 210, 215, 216, 217
policy makers, 169, 177, 216, 217
policy making, 169
policy options, 172
pollutants, 165, 175, 179, 181, 188, 192, 195, 196, 198, 202, 206, 207, 218
polluters, 206
pollution, 106, 167, 168, 171, 172, 174, 178, 179, 180, 182, 187, 188, 200, 201, 202, 205, 206, 207
polychlorinated biphenyl, 157
polycyclic aromatic hydrocarbon, 175
polymer, 101, 210
polynomial functions, 12
polypropylene, 128
polyurethane, 101
polyvinylchloride, 174
population, 111, 206
population growth, 206
Port of Koper, 207, 210, 215
portfolio, 86, 99
Portugal, 165
postal service, 87
precipitation, 195, 198
premature death, 188
preparation, 158, 162, 164
preparedness, 162
prevention, 52, 123, 155, 206
principles, 5, 187, 189, 210, 217
probability, 89, 140
process control, 65
procurement, 208
product design, 86
programming, 96
project, 48, 164, 187, 188, 189, 190, 193, 198, 201, 215, 218
propagation, 116
propane, 119
prosperity, 206
protection, 119, 157, 158, 160, 162, 163, 200, 206
prototype(s), 42, 81, 165
public policy, 216
pulmonary diseases, 188
pumps, 50, 54, 56, 57, 58, 59, 190
PVC, 174

Q

quality of life, 167
quality standards, 187, 188
quantification, 215, 216, 218
query, 88
questionnaire, 207

R

racing, 111
radar, 100
radiation, 93, 94, 95, 115, 116, 119
radio, 87
radius, 25, 111, 139
range of flexibility, 6
ratification, 160, 168
raw materials, 160, 189, 194
reaction rate, 117
reactions, 56
reading, 145
reality, 217
reasoning, 86, 100
reception, 217
receptors, 195, 196, 197
recognition, 82
recommendations, 178, 207, 217
reconstruction, 182
recovery, 169, 208, 211, 212, 213
recreational, 5, 44, 206
recycling, 155, 156, 159, 161, 162, 163, 164, 165, 166, 169, 208, 211
redundancy, 61
reference system, 189, 193, 202
regression, 7, 67, 68, 79, 83
regression method, 64, 66
regression model, 63, 64, 69
regulations, 48, 105, 106, 111, 115, 161, 164, 169, 171, 172, 173, 174, 176, 177, 200
regulatory requirements, 188
rejection, 76, 134, 140
relaxation, 101
reliability, 64
relief, 55, 57, 157
repair, 99, 169
representativeness, 178
requirements, 9, 22, 27, 29, 43, 49, 50, 131, 160, 161, 162, 166, 171, 173, 188, 190, 205, 217
researchers, 49, 64, 125
reserves, 57
residuals, 15, 193, 194
residues, 211

226 Index

resistance, 107, 120, 127, 128, 132
resolution, 173, 176, 178
resources, 81, 207, 208, 210, 213, 215, 216
response, 77, 80, 81, 83, 88, 157, 159
restoration, 99
restrictions, 77, 167
retirement, 146, 149
risk(s), vii, 108, 109, 110, 114, 115, 116, 120, 123, 133, 157, 159, 208
risk assessment, 115
rods, 53
routes, 170, 182, 205
rubber, 157
rules, 66, 72, 82, 85, 86, 99, 115, 160, 162, 163, 168, 169, 173, 217
Russia, 180

S

safety, 48, 51, 61, 106, 108, 109, 111, 114, 117, 118, 119, 123, 124, 125, 134, 146, 149, 150, 156, 159, 160, 162, 163, 164, 173, 174, 210
saturation, 58, 77
scaling, 8, 121, 172
scope, 86, 119, 125, 208, 209
sea-level, 195, 198
security, 105, 106, 109, 110, 115, 122, 123, 125, 173
semiconductor, 101
semiconductor lasers, 101
sensors, 54, 55, 146, 148
services, 51, 52, 53, 55, 58, 110, 157, 189, 206, 207, 208, 210, 216
sewage, 205
shape, 8, 9, 12, 20, 21, 26, 27, 33, 34, 35, 39, 120, 127, 130, 137, 200
ship design, vii, 6, 7, 9, 12, 22, 23, 47
shock, 110, 115, 120, 134
shock waves, 115
shoreline, 192, 196
showing, 80
signals, 51, 94, 147, 148
signs, 146
silver, 157
simulation(s), 65, 71, 79, 80, 81, 85, 86, 100, 187, 188, 189, 195, 196, 198, 199, 200
sludge, 206, 210
smog, 188
software, 8, 9, 35, 211
solid waste, 211, 212, 213
solution, 8, 10, 12, 14, 15, 21, 25, 28, 30, 33, 34, 39, 63, 64, 67, 81, 192, 193, 197, 205
South Korea, 159
sovereignty, 170, 174, 206

Spain, 5, 47, 63, 105, 114, 116, 125, 155, 159, 160, 167, 201, 206
species, 175
specifications, 50, 64, 65, 66, 83, 96, 127, 146, 173
speed of light, 88
stability, 5, 6, 7, 9, 27, 35, 38, 44, 195, 198
stakeholders, 216
state(s), 29, 54, 56, 69, 73, 74, 78, 156, 160, 161, 162, 164, 169, 170, 173, 174, 177, 181, 193, 195, 198
steel, 109, 125, 126, 127, 128, 130, 132, 135, 139, 145, 146, 148, 157, 159, 165
storage, 106, 107, 108, 109, 111, 118, 121, 123, 191
stress, 137, 139
stroke, 61, 172, 173, 175
structure, 71, 78, 86, 120, 156, 168
substitution, 18, 21, 69
sulphur, 105, 171, 172, 174, 175, 178, 179, 180, 194, 212
suppliers, 174
supply chain, 206
suppression, 179
surface area, 128
surface layer, 170
surplus, 213
sustainability, 201, 208, 216, 217
sustainable development, 182, 208, 210
Sweden, 201
Switzerland, 160
synthetic fiber, 132

T

Taiwan, 202, 218
tangent angle, 5, 8, 9, 10, 17, 18, 19, 25, 27, 34
tanks, 53, 57, 58, 83, 106, 108, 109, 111, 118
target, 8
techniques, 7, 8, 9, 43, 45, 63, 64, 82, 145
technology(ies), 61, 82, 123, 155, 165, 173, 177, 180, 182, 183, 215
temperature, 47, 50, 51, 52, 53, 55, 57, 59, 117, 119, 195, 198
tensile strength, 128
terminals, 173, 207, 210
territory, 87, 202, 216
testing, 86, 146, 147, 148
textbooks, 9
thin films, 101
time resolution, 176
time series, 198
topology, 10, 37, 44, 64
toxic substances, 159, 164
toxic waste, 160, 164

Index

227

toxicity, 209, 212, 213, 214
trade, 48, 67, 166, 167, 176, 206, 208, 210
trade costs, 176
training, 67, 69, 79, 157, 159, 162, 192
transformation(s), 8, 43, 170, 173, 181, 195, 212
transmission, 90, 117, 119
transport, 122, 166, 167, 168, 169, 176, 181, 188, 193, 194, 195, 200, 202, 205, 206, 207, 208, 210, 211, 215, 216, 217
transportation, 106, 119, 169, 172, 177, 180, 182, 187, 201, 206
treaties, 155
treatment, 53, 99, 156, 164, 165, 172, 205, 207, 211, 212, 213
trial, 6, 8
turbulence, 195
Turkey, 171, 180, 201

U

UK, 44, 202
UNFCCC, 176, 182
unforeseen circumstances, 110
uniform, 12, 13, 20, 25, 81, 127, 130, 192, 198
United Nations (UN), 156, 166, 168, 179, 208, 218
United States, 87
updating, 164
urban, 177, 188, 201, 202, 216
urban areas, 188

V

vacuum, 55, 71
validation, 65
valuation, 194
valve, 50, 51, 52, 53, 54, 55, 56, 58, 59, 117
vanadium, 175
variables, 20, 69, 87, 96
variations, 17, 148, 172
vector, 12, 13, 20, 25, 67, 68, 69, 83, 96, 97, 98
vegetation, 200
vehicles, 114, 188, 210
velocity, 69, 70, 71, 72, 80, 81, 147
Venezuela, 111, 113
ventilation, 189
vessels, 5, 7, 44, 47, 48, 60, 61, 106, 108, 109, 115, 123, 125, 126, 155, 156, 159, 161, 162, 164, 165,

168, 171, 173, 174, 175, 178, 180, 188, 191, 192, 194, 196, 206, 211, 215
vibration, 144
voiding, 53
volatile organic compounds, 173, 206

W

war, 123
Washington, 218
waste, 106, 156, 159, 160, 164, 165, 171, 174, 205, 206, 207, 210, 211, 212, 213, 215, 216, 217, 218
waste disposal, 206, 207
waste management, 205, 207, 210, 216, 218
waste treatment, 211, 212, 213
wastewater, 211
water, 9, 10, 11, 12, 20, 27, 29, 34, 37, 41, 42, 43, 47, 50, 51, 52, 53, 55, 56, 57, 58, 59, 60, 116, 117, 119, 157, 159, 165, 178, 191, 202, 205, 206, 207, 208, 211, 212
water heater, 55, 56, 57, 58
water quality, 207
weakness, 118
wealth, 109
wear, 127, 128, 133, 134, 135, 136, 137, 139, 141, 146, 149
web, 192
wind power, 206
wind speed, 195, 198
windows, 111
wires, 126, 127, 128, 129, 130, 131, 132, 133, 134, 135, 136, 139, 140, 142, 143, 144, 145, 146, 147, 148
wood, 42
workers, 157, 159, 160, 163
working conditions, 110, 157, 208
worldwide, 187

X

X-axis, 13, 22, 23, 31, 37

Z

zinc, 133, 212